## DO YOU KNOW WHAT'S IN WHAT YOU EAT?

What's the real difference—nutritionally and calorically—between buttermilk and yogurt?

Which is more caloric—sherbet or ice milk?

Which is healthier—cow's milk or goat's milk?

How lean is lean roast beef?

Do you know the nutritional differences between dark and light meat turkey?

Is there a reason why you should choose small rather than large curd creamed cottage cheese?

Which have better nutritional value—red raspberries or black raspberries?

Whether you're watching calories or carbohydrates, sodium or fats, or taking care of your family's nutritional needs, the answers to all of these questions and more are readily available in this book—your one guide to good eating, good health and true economy.

**WHAT'S IN WHAT YOU EAT**

# WHAT'S IN WHAT YOU EAT

## Condensed from the U.S. Dept. Agriculture Nutritive Value of American Foods

Designed by Will Eisner Studios
Edited by Catherine Yronwode
Illustrations by Robert Pizzo

*Produced by Poorhouse Press*

BANTAM BOOKS
TORONTO · NEW YORK · LONDON · SYDNEY

WHAT'S IN WHAT YOU EAT
*A Bantam Book / March 1983*

*The format and concept of this book is designed to provide a unique
easy-to-use handy reference for the average consumer. The
application of illustrations and graphic symbols were devised
by the editors to enhance readability and easy access.*

*All the information appears in the Government's booklet* Agriculture
Handbook No. 456 *(issued Nov. 1975 and revisions April 1977,
September 1978 in Bulletin No. 72) which was originally published
by its Agriculture Research Service.*

*Edited by Cat Yronwode*
*Illustrations by Robert Pizzo*
*Designed by Eisner Studios*

*All rights reserved.*
*Copyright © 1983 by Poorhouse Press.*
*Cover photo copyright © 1983 by Bantam Books, Inc.*
*This book may not be reproduced in whole or in part, by
mimeograph or any other means, without permission.*
*For information address: Bantam Books, Inc.*

ISBN 0-553-20925-6

*Published simultaneously in the United States and Canada*

*Bantam Books are published by Bantam Books, Inc. Its trademark,
consisting of the words "Bantam Books" and the portrayal of a
rooster, is Registered in U.S. Patent and Trademark Office and in
other countries. Marca Registrada. Bantam Books, Inc., 666 Fifth
Avenue, New York, New York 10103.*

PRINTED IN THE UNITED STATES OF AMERICA

O     0 9 8 7 6 5 4 3 2 1

# INTRODUCTION

At no time in America's history has there been greater interest in the twin subjects of nutrition and health. With a large segment of the population on one kind of diet or another, and an almost equally large portion engaged in such health-building endeavors as jogging and exercise programs, it seems that we have become a nation of nutrition and health experts. At the core of this widespread movement is the simple desire to feel good. Most of us understand that if we get the nutrients we need from the foods we eat, we will feel more energetic and have greater resistance to colds and infections.

Many maladies of our times, such as heart disease, schizophrenia, and arteriosclerosis have been tentatively linked to deficiencies in the American diet by certain researchers, and this work has been widely popularized. If so many Americans are health-conscious to this extent, why then do these problems arise at all? The answer may lie with the fact that the industrial revolution, along with its many improvements in our standard of living, brought us the practice of processing our foods in many ways to make them easier to transport and quicker to prepare for the table. Fifty years ago, the food we consumed was generally fresh and unprocessed. Now many of our foods are canned, pre-cooked, or shipped long distances, losing nutritional value each step of the way.

What does this mean for the average consumer? It means that we must become even more aware of the nutritional value of the foods we buy. It is more important than ever to choose our foods carefully. This is not to say that all processed foods are lacking in nutritional value, only that we have to go beyond the nutritional basics we learned in grade school if we are to get the most for our money and do the most for our bodies. And yet who has time to take on the project of gleaning through many obscure and scientifically oriented volumes to determine just what nutrients are to be found in the foods we most often consume? For those who want the facts, without the problem of digging them out of nutritional textbooks and encyclopedias, this book, which presents the nutritional value of our most common foods, in the forms we most commonly find them — and does so in an easily readable way — can be a valuable and convenient adjunct to a program of good nutrition.

It doesn't take a dietician's approach to meal planning to improve the health and well being of our families and ourselves. Great changes in diet are not as important to the average American as learning how to get the most food value from those things we like best to eat. In this way we can be assured that we are obtaining all the vitamins, minerals and protein which our bodies need, while still enjoying our meals to the fullest.

In addition to choosing a balanced diet of healthful foods, it is good common sense to use good cooking practices and avoid destroying or throwing out vitamins which are heat sensitive and water soluble, such as Vitamin C. Vegetables boiled in large amounts of water and drained before serving lose a great deal of their original vitamin and mineral content, whereas vegetables steamed in a small amount of water are not only attractive, they are actually more nutritious. Some of the best soups are those made with a ''stock'' composed of cooking water saved from previously cooked vegetables and meats. Not only does the ''stock'' add delicious flavor, it reclaims vital nutrients which would otherwise go down the drain. Another practice, which maintains both taste and nutritional value in fresh fruits and vegetables, is to cut them immediately before eating, if they are to be cut at all. Leaving cut fruits and vegetables in the refrigerator leads to a loss of Vitamin C.

These simple cooking tips, combined with the easy-to-understand information charts in this book, can be your key to getting the most out of your food dollar and enabling your body to make the best possible use of what you eat. My hope in presenting this book is that good health and happiness will be within the reach of us all.

Catherine Yronwode
June, 1982

# DICTIONARY OF COMMON NUTRIENTS

**Ascorbic Acid (Vitamin C):** Helps heal wounds, maintains collagen (in connective tissue and bones), strengthens blood vessels and helps the body absorb iron. It may also provide resistance to some forms of infection.

**Calcium:** Helps develop and maintain strong bones and teeth. Aids in the normal clotting of blood, muscular response and the functioning of the heart and nerves.

**Calorie:** A unit used in the measurement of heat and energy.

**Fat:** Provides energy and acts as a base for the fat-soluble vitamins (A, D, E, K). It also supplies fatty acids, essential for growth and health, as well as smooth skin.

**Iron:** Is essential in the production of the blood components hemoglobin and myoglobin. Also promotes growth and aids in the metabolism of protein.

**Niacin (Nicotinic Acid, Niacinamide):** Is essential in the metabolism of protein, fat and carbohydrates. Also aids in the maintenance of healthy skin, digestive system and tongue.

**Phosphorus:** In conjunction with calcium, works to build bones and teeth.

**Potassium:** Necessary for regulation of heart muscles, nervous system and kidneys.

**Protein:** A source of heat and energy which is also necessary for growth and development of the body. It is essential to the formation of enzymes, hormones and antibodies, and maintains the acid-alkali balance of the system.

**Sodium:**   Necessary in the regulation of fluid levels in cells. Maintains normal function and health of the nervous system, lymph system, muscles and blood system.

**Thiamin (Vitamin B₁):**   Essential to the metabolism of carbohydrates. Aids in the maintenance of the nervous system. Stimulates growth and development of muscle tone. Stabilizes the appetite.

**Vitamin A (Carotene):**   Aids in the formation of bones and teeth and is essential in the growth and repair of all body tissues. Important to the health and functioning of the eyes. Aids the maintenance of healthy epithelial tissue and combats bacteria and other infections.

**Water:**   Utilized in almost every body function, including digestion, excretion, absorption and circulation. It is the carrier of nutrients throughout the body, helps carry waste material out of the body and aids in the maintenance of normal body temperature.

# MILK AND
# MILK PRODUCTS

BUTTERMILK

CHOCOLATE MILK (COMMERCIAL)

COCOA AND HOT CHOCOLATE (HOMEMADE)

CUSTARD

DRIED MILK

EGGNOG (COMMERCIAL)

ICE CREAM, HARDENED, REGULAR
(11% BUTTER FAT)

ICE CREAM, HARDENED, RICH
(16% BUTTER FAT)

ICE CREAM, SOFT, REGULAR
(11% BUTTER FAT)

ICE MILK, HARDENED (4.3% BUTTER FAT)

ICE MILK, SOFT (2.6% BUTTER FAT)

LOWFAT MILK (1% BUTTER FAT)

LOWFAT MILK (2% BUTTER FAT)

MALTED MILK

MILK SHAKES

NONFAT (SKIM) MILK

PUDDINGS, FROM MIX

PUDDINGS, HOMEMADE

SHERBERT

SWEETENED CONDENSED MILK

UNSWEETENED EVAPORATED MILK

WHOLE MILK, COW

WHOLE MILK, GOAT

YOGURT

# BUTTERMILK

| | grams | | |
|---|---|---|---|
| **Weight:** | | 245 | 980 |

| | | | |
|---|---|---:|---:|
| % | **Water** | 90 | 90 |
| | **Calories** | 100 | 400 |
| grams | **Protein** | 8 | 33 |
| | **Fat** | .2 | 1 |
| | **Carbohydrate** | 12 | 50 |
| | **Calcium** | 285 | 1,186 |
| | **Phosphorus** | 219 | 930 |
| | **Iron** | .1 | .4 |
| milligrams | **Sodium** | 319 | 1,274 |
| | **Potassium** | 371 | 1,374 |
| | **Thiamin** | .08 | .33 |
| | **Riboflavin** | .38 | 1.58 |
| | **Niacin** | .1 | .5 |
| | **Ascorbic Acid** | 2 | 10 |
| unit | **Vitamin A** | *80 | *320 |

*Applies to product without Vitamin A added.

4

# CHOCOLATE MILK
## (COMMERCIAL)

| | | | | |
|---|---|---|---|---|
| grams | **Weight:** | 250 | 250 | 250 |

| | | Regular | Low fat (2%) | Low fat (1%) |
|---|---|---|---|---|
| % | **Water** | 82 | 84 | 85 |
| | **Calories** | 210 | 180 | 160 |
| grams | **Protein** | 8 | 8 | 8 |
| | **Fat** | 8 | 5 | 3 |
| | **Carbohydrate** | 26 | 26 | 26 |
| | **Calcium** | 280 | 284 | 287 |
| milligrams | **Phosphorus** | 251 | 254 | 257 |
| | **Iron** | .6 | .6 | .6 |
| | **Sodium** | 118 | 115 | 120 |
| | **Potassium** | 417 | 422 | 426 |
| | **Thiamin** | .09 | .10 | .10 |
| | **Riboflavin** | .41 | .42 | .40 |
| | **Niacin** | .3 | .3 | .2 |
| | **Ascorbic Acid** | 2 | 2 | 2 |
| unit | **Vitamin A** | *300 | 500 | 500 |

*Applies to product without Vitamin A added.

5

# COCOA AND
# HOT CHOCOLATE
## HOMEMADE

| grams | **Weight:** | 250 | 250 |
|---|---|---|---|

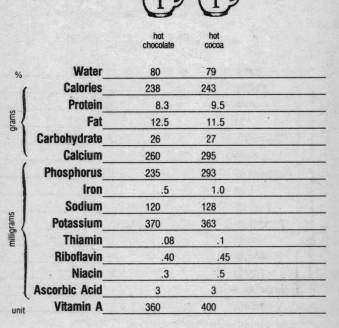

| | | hot chocolate | hot cocoa |
|---|---|---|---|
| % | **Water** | 80 | 79 |
| | **Calories** | 238 | 243 |
| grams | **Protein** | 8.3 | 9.5 |
| | **Fat** | 12.5 | 11.5 |
| | **Carbohydrate** | 26 | 27 |
| | **Calcium** | 260 | 295 |
| milligrams | **Phosphorus** | 235 | 293 |
| | **Iron** | .5 | 1.0 |
| | **Sodium** | 120 | 128 |
| | **Potassium** | 370 | 363 |
| | **Thiamin** | .08 | .1 |
| | **Riboflavin** | .40 | .45 |
| | **Niacin** | .3 | .5 |
| | **Ascorbic Acid** | 3 | 3 |
| unit | **Vitamin A** | 360 | 400 |

6

# CUSTARD
## BAKED

grams **Weight:** 265

| | | | |
|---|---|---|---|
| % | **Water** | 77 | |
| | **Calories** | 305 | |
| grams | **Protein** | 14 | |
| | **Fat** | 15 | |
| | **Carbohydrate** | 29 | |
| | **Calcium** | 297 | |
| | **Phosphorus** | 310 | |
| | **Iron** | 1.1 | |
| milligrams | **Sodium** | 209 | |
| | **Potassium** | 387 | |
| | **Thiamin** | .11 | |
| | **Riboflavin** | .50 | |
| | **Niacin** | .3 | |
| | **Ascorbic Acid** | 1 | |
| unit | **Vitamin A** | 930 | |

# DRIED MILK
# (AND DRIED BUTTERMILK)

| | Weight: | 120 | 91 | 68 |
|---|---|---|---|---|
| grams | | | | |
| | | Buttermilk | nonfat instant (envelope net wt.) 3.2 oz. | regular instant cup**** |
| % | Water | 3 | 4 | 4 |
| | Calories | 465 | 325 | 245 |
| grams | Protein | 41 | .32 | 24 |
| | Fat | 7 | 1 | Trace |
| | Carbohydrate | 59 | 47 | 35 |
| | Calcium | 1,421 | 1,120 | 837 |
| milligrams | Phosphorus | 1,119 | 896 | 670 |
| | Iron | .4 | .3 | .2 |
| | Sodium | 608 | 479 | 358 |
| | Potassium | 1,910 | 1,552 | 1,160 |
| | Thiamin | .47 | .38 | .28 |
| | Riboflavin | 1.90 | 1.59 | 1.19 |
| | Niacin | 1.1 | .8 | .6 |
| | Ascorbic Acid | 7 | 5 | 4 |
| unit | Vitamin A | *260 | **2,160 | ***1,610 |

*Applies to product without Vitamin A added.    ***Vitamin A added.
**Yields 1 qt. of fluid milk when reconstituted according to package directions.
****Weight applies to product with label claim of 1⅓ cups equal 3.2 oz.

# EGGNOG
## (COMMERCIAL)

grams    **Weight:**     254

| | | |
|---|---|---|
| % | Water | 74 |
| | Calories | 340 |
| grams | Protein | 10 |
| | Fat | 19 |
| | Carbohydrate | 34 |
| | Calcium | 330 |
| milligrams | Phosphorus | 278 |
| | Iron | .5 |
| | Sodium | 312 |
| | Potassium | 420 |
| | Thiamin | .09 |
| | Riboflavin | .48 |
| | Niacin | .3 |
| | Ascorbic Acid | 4 |
| unit | Vitamin A | 890 |

9

# ICE CREAM
## HARDENED, REGULAR

| grams | **Weight:** | 1,064 | 133 | 50 |
|---|---|---|---|---|

| | | | | |
|---|---|---|---|---|
| % | **Water** | 61 | 61 | 61 |
| | **Calories** | 2,155 | 270 | 100 |
| grams | **Protein** | 38 | 5 | 2 |
| | **Fat** | 115 | 14 | .5 |
| | **Carbohydrate** | 254 | 32 | 12 |
| | **Calcium** | 1,406 | 176 | 66 |
| milligrams | **Phosphorus** | 1,075 | 134 | 51 |
| | **Iron** | 1.0 | .1 | Trace |
| | **Sodium** | 670 | 84 | N.A. |
| | **Potassium** | 2,052 | 257 | 96 |
| | **Thiamin** | .42 | .05 | .02 |
| | **Riboflavin** | 2.63 | .33 | .12 |
| | **Niacin** | 1.1 | .1 | .1 |
| | **Ascorbic Acid** | 6 | 1 | Trace |
| unit | **Vitamin A** | 4,340 | 540 | 200 |

*About 11% fat.

# ICE CREAM
## HARDENED, RICH

grams **Weight:** 1,188 148

\* \*

| | | |
|---|---|---|
| % | **Water** | 59 | 59 |
| | **Calories** | 2,805 | 350 |
| grams | **Protein** | 33 | 4 |
| | **Fat** | 190 | 24 |
| | **Carbohydrate** | 256 | 32 |
| | **Calcium** | 1,213 | 151 |
| | **Phosphorus** | 927 | 115 |
| | **Iron** | .8 | .1 |
| milligrams | **Sodium** | 392 | 49 |
| | **Potassium** | 1,771 | 221 |
| | **Thiamin** | .36 | .04 |
| | **Riboflavin** | 2.27 | .28 |
| | **Niacin** | .9 | .1 |
| | **Ascorbic Acid** | 5 | 1 |
| unit | **Vitamin A** | 7,200 | 900 |

*About 16% fat.

11

# SOFT, REGULAR
# ICE CREAM
## (FROZEN CUSTARD)

grams    **Weight:**          173

*

| | | |
|---|---:|---|
| %    **Water** | 60 | |
| **Calories** | 375 | |
| **Protein** | 7 | |
| **Fat** | 23 | |
| **Carbohydrate** | 38 | |
| **Calcium** | 236 | |
| **Phosphorus** | 199 | |
| **Iron** | .4 | |
| **Sodium** | 119 | |
| **Potassium** | 338 | |
| **Thiamin** | .08 | |
| **Riboflavin** | .45 | |
| **Niacin** | .2 | |
| **Ascorbic Acid** | 1 | |
| unit    **Vitamin A** | 790 | |

grams { (Calories, Protein, Fat, Carbohydrate, Calcium)

milligrams { (Phosphorus, Iron, Sodium, Potassium, Thiamin, Riboflavin, Niacin, Ascorbic Acid)

*About 11% fat.

12

# ICE MILK
## HARDENED

| | | |
|---|---|---|
| grams | **Weight:** | 1,048 | 131 |

| | | | |
|---|---|---|---|
| % | **Water** | 69 | 69 |
| | **Calories** | 1,470 | 185 |
| grams | **Protein** | 41 | 5 |
| | **Fat** | 45 | 6 |
| | **Carbohydrate** | 232 | 29 |
| | **Calcium** | 1,409 | 176 |
| milligrams | **Phosphorus** | 1,035 | 129 |
| | **Iron** | 1.5 | .1 |
| | **Sodium** | 713 | 89 |
| | **Potassium** | 2,117 | 265 |
| | **Thiamin** | .61 | .08 |
| | **Riboflavin** | 2.78 | .35 |
| | **Niacin** | .9 | .1 |
| | **Ascorbic Acid** | 6 | 1 |
| unit | **Vitamin A** | 1,710 | 210 |

*About 4.3% fat.

13

# ICE MILK
## SOFT

| | Weight: | 175 |
|---|---|---|
| grams | | |

\*

| | | |
|---|---|---|
| % | **Water** | 70 |
| | **Calories** | 225 |
| | **Protein** | 8 |
| grams | **Fat** | 5 |
| | **Carbohydrate** | .38 |
| | **Calcium** | 274 |
| | **Phosphorus** | 202 |
| | **Iron** | .3 |
| | **Sodium** | 119 |
| | **Potassium** | 412 |
| milligrams | **Thiamin** | .12 |
| | **Riboflavin** | .54 |
| | **Niacin** | .2 |
| | **Ascorbic Acid** | 1 |
| unit | **Vitamin A** | 180 |

\*About 2.6% fat.

# LOWFAT MILK
# (1%)

| | | no milk solids added | milk solids added (less than 10g protein) | milk solids added: more than 10g protein |
|---|---|---|---|---|
| grams | **Weight:** | 244 | 245 | 246 |
| % | **Water** | 90 | 90 | 89 |
| | **Calories** | 100 | 105 | 120 |
| grams | **Protein** | 8 | 9 | 10 |
| | **Fat** | 3 | 2 | 3 |
| | **Carbohydrate** | 12 | 12 | 14 |
| | **Calcium** | 300 | 313 | 349 |
| milligrams | **Phosphorus** | 235 | 245 | 273 |
| | **Iron** | .1 | .1 | .1 |
| | **Sodium** | N.A. | N.A. | N.A. |
| | **Potassium** | 381 | 397 | 444 |
| | **Thiamin** | .10 | .10 | .11 |
| | **Riboflavin** | .41 | .42 | .47 |
| | **Niacin** | .2 | .2 | .2 |
| | **Ascorbic Acid** | 2 | 2 | 3 |
| unit | **Vitamin A** | 50 | 500 | 500 |

# LOWFAT MILK
## (2% FAT)

| | | | |
|---|---|---|---|
| grams | **Weight:** | 244 | 245 | 246 |

| | | no milk solids added | milk solids added * | milk solids added ** |
|---|---|---|---|---|
| % | **Water** | 89 | 89 | 88 |
| | **Calories** | 120 | 125 | 135 |
| grams | **Protein** | 8 | 9 | 10 |
| | **Fat** | 5 | 5 | 5 |
| | **Carbohydrate** | 12 | 12 | 14 |
| | **Calcium** | 297 | 313 | 352 |
| milligrams | **Phosphorus** | 232 | 245 | 276 |
| | **Iron** | .1 | .1 | .1 |
| | **Sodium** | 127 | 150 | 150 |
| | **Potassium** | 377 | 397 | 447 |
| | **Thiamin** | .10 | .10 | .11 |
| | **Riboflavin** | .40 | .42 | .48 |
| | **Niacin** | .2 | .2 | .2 |
| | **Ascorbic Acid** | 2 | 2 | 3 |
| unit | **Vitamin A** | 500 | 500 | 500 |

*Label claim less than 10g protein per cup.
**More than 10g protein per cup.

16

# MALTED MILK
## BEVERAGES, HOMEMADE

| | grams | Weight: | 265 | 265 |
|---|---|---|---|---|

chocolate*     natural*

| | | chocolate* | natural* |
|---|---|---|---|
| % | Water | 81 | 81 |
| | Calories | 235 | 235 |
| grams { | Protein | 9 | 11 |
| | Fat | 9 | 10 |
| | Carbohydrate | 29 | 27 |
| | Calcium | 304 | 347 |
| milligrams { | Phosphorus | 265 | 307 |
| | Iron | .5 | .3 |
| | Sodium | 214 | 214 |
| | Potassium | 500 | 529 |
| | Thiamin | .14 | .20 |
| | Riboflavin | .43 | .54 |
| | Niacin | .7 | 1.3 |
| | Ascorbic Acid | 2 | 2 |
| unit | Vitamin A | 330 | 380 |

*Home-prepared with 1 cup of whole milk and 2-3 heaping tsp. of malted milk powder (about ¾ oz.)

17

# MILK SHAKES
## THICK*

| | grams | **Weight:** | 300 | 313 |

chocolate container net wt. 10.6 oz.     vanilla container net wt. 11 oz.

| | | chocolate | vanilla |
|---|---|---|---|
| % | **Water** | 72 | 74 |
| grams | **Calories** | 355 | 350 |
| | **Protein** | 9 | 12 |
| | **Fat** | 8 | 9 |
| | **Carbohydrate** | 63 | 56 |
| | **Calcium** | 396 | 457 |
| milligrams | **Phosphorus** | 378 | 361 |
| | **Iron** | .9 | .3 |
| | **Sodium** | N.A. | N.A. |
| | **Potassium** | 672 | 572 |
| | **Thiamin** | .14 | .09 |
| | **Riboflavin** | .67 | .61 |
| | **Niacin** | .4 | .5 |
| | **Ascorbic Acid** | 0 | 0 |
| unit | **Vitamin A** | 260 | 360 |

*Applies to products made from thick shake mixes and that do not contain added ice cream. Products made from milk shake mixes are higher in fat and usually contain added ice cream.

18

# NONFAT
# SKIM MILK

| | | | |
|---|---|---|---|
| grams | **Weight:** | 245 | 245 | 246 |

| | | no milk solids added | milk solids added (label claims less than 10g protein per cup) | milk solids added (label claims more than 10g protein per cup) |
|---|---|---|---|---|
| % | **Water** | 91 | 90 | 89 |
| | **Calories** | 85 | 90 | 100 |
| grams | **Protein** | 8 | 9 | 10 |
| | **Fat** | Trace | 1 | 1 |
| | **Carbohydrate** | 12 | 12 | 14 |
| | **Calcium** | 302 | 316 | 352 |
| milligrams | **Phosphorus** | 247 | 255 | 275 |
| | **Iron** | .1 | .1 | .1 |
| | **Sodium** | 127 | N.A. | N.A. |
| | **Potassium** | 406 | 418 | 446 |
| | **Thiamin** | .09 | .10 | .11 |
| | **Riboflavin** | .37 | .43 | .48 |
| | **Niacin** | .2 | .2 | .2 |
| | **Ascorbic Acid** | 2 | 2 | 3 |
| unit | **Vitamin A** | 500 | 500 | 500 |

# PUDDINGS
## FROM MIX

| | | Regular (cooked) | Instant |
|---|---|---|---|
| grams | **Weight:** | 260 | 260 |

| | | Regular (cooked) | Instant |
|---|---|---|---|
| % | **Water** | 70 | 69 |
| | **Calories** | 320 | 325 |
| | **Protein** | 9 | 8 |
| | **Fat** | 8 | 7 |
| | **Carbohydrate** | 59 | 63 |
| | **Calcium** | 265 | 374 |
| | **Phosphorus** | 247 | 237 |
| | **Iron** | .8 | 1.3 |
| | **Sodium** | 335 | 322 |
| | **Potassium** | 354 | 335 |
| | **Thiamin** | .05 | .08 |
| | **Riboflavin** | .39 | .39 |
| | **Niacin** | .3 | .3 |
| | **Ascorbic Acid** | 2 | 2 |
| unit | **Vitamin A** | 340 | 340 |

(grams: Protein, Fat, Carbohydrate; milligrams: Calcium, Phosphorus, Iron, Sodium, Potassium, Thiamin, Riboflavin, Niacin, Ascorbic Acid)

# PUDDINGS
## HOME RECIPE

| grams | Weight: | 260 | 255 | 165 |
|---|---|---|---|---|

| | | Chocolate starch base | Vanilla* starch base | Tapioca Cream |
|---|---|---|---|---|
| % | **Water** | 66 | 76 | 165 |
| | **Calories** | 385 | 285 | 220 |
| grams | **Protein** | 8 | 9 | 8 |
| | **Fat** | 12 | 10 | 8 |
| | **Carbohydrate** | 67 | 41 | 28 |
| | **Calcium** | 250 | 298 | 173 |
| milligrams | **Phosphorus** | 255 | 232 | 180 |
| | **Iron** | 1.3 | Trace | .7 |
| | **Sodium** | 146 | 166 | 257 |
| | **Potassium** | 445 | 352 | 223 |
| | **Thiamin** | .05 | .08 | .07 |
| | **Riboflavin** | .36 | .41 | .30 |
| | **Niacin** | .3 | .3 | .2 |
| | **Ascorbic Acid** | 1 | 2 | 2 |
| unit | **Vitamin A** | 390 | 410 | 480 |

*Blanc mange.

21

# SHERBERT

| grams | Weight: | 1,542 | 193 |
|---|---|---|---|

|  |  |  |  |
|---|---|---|---|
| % | **Water** | 66 | 66 |
|  | **Calories** | 2,160 | 270 |
| grams | **Protein** | 17 | 2 |
|  | **Fat** | 31 | 4 |
|  | **Carbohydrate** | 469 | 59 |
|  | **Calcium** | 827 | 103 |
| milligrams | **Phosphorus** | 594 | 74 |
|  | **Iron** | 2.5 | 3 |
|  | **Sodium** | 22 | 1 |
|  | **Potassium** | 1,585 | 198 |
|  | **Thiamin** | .26 | .03 |
|  | **Riboflavin** | .71 | .09 |
|  | **Niacin** | 1.0 | .1 |
|  | **Ascorbic Acid** | 31 | 4 |
| unit | **Vitamin A** | 1,480 | 190 |

*About 2% fat.

# SWEETENED CONDENSED
# MILK

grams   **Weight:**    306

| | | |
|---|---|---|
| %   **Water** | 27 | |
| **Calories** | 980 | |
| **Protein** | 24 | |
| **Fat** | 27 | |
| **Carbohydrate** | 166 | |
| **Calcium** | 868 | |
| **Phosphorus** | 775 | |
| **Iron** | .6 | |
| **Sodium** | 343 | |
| **Potassium** | 1,136 | |
| **Thiamin** | .28 | |
| **Riboflavin** | 1.27 | |
| **Niacin** | .6 | |
| **Ascorbic Acid** | 8 | |
| unit   **Vitamin A** | *1,000 | |

grams (Calories–Calcium); milligrams (Phosphorus–Vitamin A)

*Applies to product without Vitamin A added.

# UNSWEETENED EVAPORATED
# MILK

| | | |
|---|---|---|
| grams | **Weight:** | 252 | 255 |

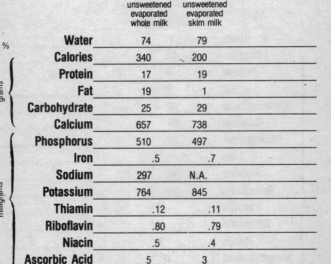

| | | unsweetened evaporated whole milk | unsweetened evaporated skim milk |
|---|---|---|---|
| % | **Water** | 74 | 79 |
| | **Calories** | 340 | 200 |
| grams | **Protein** | 17 | 19 |
| | **Fat** | 19 | 1 |
| | **Carbohydrate** | 25 | 29 |
| | **Calcium** | 657 | 738 |
| milligrams | **Phosphorus** | 510 | 497 |
| | **Iron** | .5 | .7 |
| | **Sodium** | 297 | N.A. |
| | **Potassium** | 764 | 845 |
| | **Thiamin** | .12 | .11 |
| | **Riboflavin** | .80 | .79 |
| | **Niacin** | .5 | .4 |
| | **Ascorbic Acid** | 5 | 3 |
| unit | **Vitamin A** | *610 | **1,000 |

*Applies to product without Vitamin A added.
**Applies to product with Vitamin A added. Without added Vitamin A, value is 20 International Units (I.U.).

# WHOLE MILK
## (COW)

| grams | Weight: | 244 | 976 |
|---|---|---|---|

3.3% fat     3.3% fat

| | | | |
|---|---|---|---|
| % | **Water** | 88 | 88 |
| | **Calories** | 150 | 634 |
| grams | **Protein** | 8 | 34.2 |
| | **Fat** | 8 | 34.2 |
| | **Carbohydrate** | 11 | 47.8 |
| | **Calcium** | 291 | 1,152 |
| milligrams | **Phosphorus** | 228 | 908 |
| | **Iron** | .1 | .4 |
| | **Sodium** | 122 | 488 |
| | **Potassium** | 370 | 1,405 |
| | **Thiamin** | .09 | .29 |
| | **Riboflavin** | .40 | 1.66 |
| | **Niacin** | .2 | 1.0 |
| | **Ascorbic Acid** | 2 | 10 |
| unit | **Vitamin A** | *310 | 1,410 |

*Applies to product without Vitamin A. With added Vitamin A, value is 500
International Units (I.U.).

25

# WHOLE MILK
## (GOAT)

| | grams | | |
|---|---|---|---|
| **Weight:** | 244 | 976 | |

| | | 1 | 1 Qt. |
|---|---|---|---|

| | | | |
|---|---|---|---|
| **%** | **Water** | 87.5 | 87.5 |
| | **Calories** | 163 | 654 |
| | **Protein** | 7.8 | 31.2 |
| **grams** | **Fat** | 9.8 | 39.0 |
| | **Carbohydrate** | 11.2 | 44.9 |
| | **Calcium** | 315 | 1,250 |
| | **Phosphorus** | 259 | 1,035 |
| | **Iron** | .2 | 1.0 |
| | **Sodium** | 83 | 332 |
| **milligrams** | **Potassium** | 439 | 1,757 |
| | **Thiamin** | .10 | .39 |
| | **Riboflavin** | .27 | 1.07 |
| | **Niacin** | .7 | 2.9 |
| | **Ascorbic Acid** | 2 | 10 |
| **unit** | **Vitamin A** | 390 | 1,560 |

# YOGURT

| | | with added milk solids made with lowfat milk fruit-flavored** | with added milk solids made with lowfat milk plain | with added milk solids made with lowfat milk |
|---|---|---|---|---|
| grams | **Weight:** | 227 | 227 | 227 |
| % | **Water** | 75 | 85 | 85 |
| grams | **Calories** | 230 | 145 | 125 |
| | **Protein** | 10 | 12 | 13 |
| | **Fat** | 3 | 4 | Trace |
| | **Carbohydrate** | 42 | 16 | 17 |
| milligrams | **Calcium** | 343 | 415 | 452 |
| | **Phosphorus** | 269 | 326 | 355 |
| | **Iron** | .2 | .2 | .2 |
| | **Sodium** | N.A. | 115 | N.A. |
| | **Potassium** | 439 | 531 | 579 |
| | **Thiamin** | .08 | .10 | .11 |
| | **Riboflavin** | .40 | .49 | .53 |
| | **Niacin** | .2 | .3 | .3 |
| | **Ascorbic Acid** | 1 | 2 | 2 |
| unit | **Vitamin A** | *120 | *150 | *20 |

*Applies to product made with milk containing no added Vitamin A.
**Content of fat, Vitamin A, and carbohydrate varies. Consult the label when precise values are needed for special diets.

# YOGURT

grams    **Weight:**     227

without added
milk solids
made with
whole milk

| | % | |
|---|---|---|
| **Water** | 88 | |

| | grams | |
|---|---|---|
| **Calories** | 140 | |
| **Protein** | 8 | |
| **Fat** | 7 | |
| **Carbohydrate** | 11 | |

| | milligrams | |
|---|---|---|
| **Calcium** | 274 | |
| **Phosphorus** | 215 | |
| **Iron** | .1 | |
| **Sodium** | 125 | |
| **Potassium** | 351 | |
| **Thiamin** | .07 | |
| **Riboflavin** | .32 | |
| **Niacin** | .2 | |
| **Ascorbic Acid** | 1 | |

| | unit | |
|---|---|---|
| **Vitamin A** | 280 | |

# CREAM AND
# CREAM PRODUCTS

HALF AND HALF
LIGHT OR TABLE CREAM
PRESSURIZED WHIPPED CREAM TOPPING
SOUR CREAM
LIGHT CREAM
WHIPPING CREAM

# HALF-AND-HALF
## (CREAM AND MILK)

| | grams | Weight: | 242 | 15 |
|---|---|---|---|---|

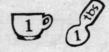

| | | | |
|---|---|---|---|
| % | **Water** | 81 | 81 |
| | **Calories** | 315 | 20 |
| grams | **Protein** | 7 | Trace |
| | **Fat** | 28 | 2 |
| | **Carbohydrate** | 10 | 1 |
| | **Calcium** | 254 | 16 |
| | **Phosphorus** | 230 | 14 |
| | **Iron** | .2 | Trace |
| | **Sodium** | 111 | 7 |
| milligrams | **Potassium** | 314 | 19 |
| | **Thiamin** | .08 | .01 |
| | **Riboflavin** | .36 | .02 |
| | **Niacin** | .2 | Trace |
| | **Ascorbic Acid** | 2 | Trace |
| unit | **Vitamin A** | 260 | 20 |

# LIGHT OR TABLE CREAM

grams     **Weight:**     240        15

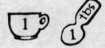

| | | |
|---|---|---|
| %    **Water** | 74 | 74 |
| **Calories** | 470 | 30 |
| **Protein** | 6 | Trace |
| **Fat** | 46 | 3 |
| **Carbohydrate** | 9 | 1 |
| **Calcium** | 231 | 14 |
| **Phosphorus** | 192 | 12 |
| **Iron** | .1 | Trace |
| **Sodium** | 103 | 6 |
| **Potassium** | 292 | 18 |
| **Thiamin** | .08 | Trace |
| **Riboflavin** | .36 | .02 |
| **Niacin** | .1 | Trace |
| **Ascorbic Acid** | 2 | Trace |
| unit    **Vitamin A** | 1,730 | 110 |

grams { Protein, Fat, Carbohydrate, Calcium

milligrams { Phosphorus ... Vitamin A

31

# PRESSURIZED
# WHIPPED CREAM
## TOPPING

| grams | **Weight:** | 60 | 3 |

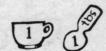

| | | | |
|---|---|---|---|
| % | **Water** | 61 | 61 |
| | **Calories** | 155 | 10 |
| grams | **Protein** | 2 | Trace |
| | **Fat** | 13 | 1 |
| | **Carbohydrate** | 7 | Trace |
| | **Calcium** | 61 | 3 |
| milligrams | **Phosphorus** | 54 | 3 |
| | **Iron** | Trace | Trace |
| | **Sodium** | N.A. | N.A. |
| | **Potassium** | 88 | 4 |
| | **Thiamin** | .02 | Trace |
| | **Riboflavin** | .04 | Trace |
| | **Niacin** | Trace | Trace |
| | **Ascorbic Acid** | 0 | 0 |
| unit | **Vitamin A** | 550 | 30 |

32

# SOUR CREAM

| | grams | Weight: | 230 | 12 |
|---|---|---|---|---|

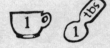

| | | | |
|---|---|---|---|
| % | Water | 71 | 71 |
| grams { | Calories | 495 | 25 |
| | Protein | 7 | Trace |
| | Fat | 48 | 3 |
| | Carbohydrate | 10 | 1 |
| | Calcium | 268 | 14 |
| milligrams { | Phosphorus | 195 | 10 |
| | Iron | .1 | Trace |
| | Sodium | N.A. | N.A. |
| | Potassium | 331 | 17 |
| | Thiamin | .08 | Trace |
| | Riboflavin | .34 | .02 |
| | Niacin | .2 | Trace |
| | Ascorbic Acid | 2 | Trace |
| unit | Vitamin A | 1,820 | 90 |

# LIGHT CREAM

| | | |
|---|---|---|
| grams | **Weight:** | 238 | 15 |

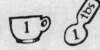

| | | | |
|---|---|---|---|
| % | **Water** | 64 | 64 |
| | **Calories** | 700 | 45 |
| | **Protein** | 5 | Trace |
| grams | **Fat** | 74 | 5 |
| | **Carbohydrate** | 7 | Trace |
| | **Calcium** | 166 | 10 |
| | **Phosphorus** | 146 | 9 |
| | **Iron** | .1 | Trace |
| | **Sodium** | 86 | 5 |
| milligrams | **Potassium** | 231 | 15 |
| | **Thiamin** | .06 | Trace |
| | **Riboflavin** | .30 | .02 |
| | **Niacin** | .1 | Trace |
| | **Ascorbic Acid** | 1 | Trace |
| unit | **Vitamin A** | 2,690 | 170 |

# HEAVY
# WHIPPING CREAM
## (UNWHIPPED)*

| | grams | | |
|---|---|---|---|
| **Weight:** | | 238 | 15 |

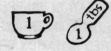

| | | | |
|---|---|---|---|
| % | **Water** | 58 | 58 |
| | **Calories** | 820 | 80 |
| grams | **Protein** | 5 | Trace |
| | **Fat** | 88 | 6 |
| | **Carbohydrate** | 7 | Trace |
| | **Calcium** | 154 | 10 |
| milligrams | **Phosphorus** | 149 | 9 |
| | **Iron** | .1 | Trace |
| | **Sodium** | 76 | 5 |
| | **Potassium** | 179 | 11 |
| | **Thiamin** | .05 | Trace |
| | **Riboflavin** | .26 | .02 |
| | **Niacin** | .1 | Trace |
| | **Ascorbic Acid** | 1 | Trace |
| unit | **Vitamin A** | 3,500 | 220 |

*Unwhipped. Volume about double when whipped.

# IMITATION CREAMERS AND CREAM PRODUCTS

FROZEN IMITATION WHIPPED TOPPING
LIQUID (FROZEN) IMITATION CREAMERS
POWDERED IMITATION CREAMERS
POWDERED IMITATION WHIPPED TOPPING
PRESSURIZED IMITATION WHIPPED TOPPING
IMITATION SOUR DRESSING

# FROZEN IMITATION
# WHIPPED TOPPING

| | grams | Weight: | 75 | 4 |
|---|---|---|---|---|

|  | made with vegetable fat | made with vegetable fat |
|---|---|---|
| % | **Water** | 50 | 50 |
| | **Calories** | 240 | 15 |
| grams | **Protein** | 1 | Trace |
| | **Fat** | 19 | 1 |
| | **Carbohydrate** | 17 | 1 |
| | **Calcium** | 5 | Trace |
| | **Phosphorus** | 6 | Trace |
| | **Iron** | .1 | Trace |
| milligrams | **Sodium** | N.A. | N.A. |
| | **Potassium** | 14 | 1 |
| | **Thiamin** | 0 | 0 |
| | **Riboflavin** | 0 | 0 |
| | **Niacin** | 0 | 0 |
| | **Ascorbic Acid** | 0 | 0 |
| unit | **Vitamin A** | *650 | *30 |

*Vitamin A value is largely from beta-carotene used for coloring.

## LIQUID (FROZEN)
# IMITATION SWEET CREAM

| grams | Weight: | 245 | 15 |
|---|---|---|---|

| | | made with vegetable fat | made with vegetable fat |
|---|---|---|---|
| % | **Water** | 77 | 77 |
| | **Calories** | 335 | 20 |
| grams | **Protein** | 2 | Trace |
| | **Fat** | 24 | 1 |
| | **Carbohydrate** | 28 | 2 |
| | **Calcium** | 23 | 1 |
| milligrams | **Phosphorus** | 157 | 10 |
| | **Iron** | .1 | Trace |
| | **Sodium** | N.A. | N.A. |
| | **Potassium** | 0 | 0 |
| | **Thiamin** | 0 | 0 |
| | **Riboflavin** | 0 | 0 |
| | **Niacin** | 0 | 0 |
| | **Ascorbic Acid** | 0 | 0 |
| unit | **Vitamin A** | *220 | *10 |

*Vitamin A value is largely from beta-carotene used for coloring. Riboflavin value for powdered creams applies to products with added riboflavin.

# POWDERED
# SWEET CREAMERS
## IMITATION

| | grams | Weight: | 94 | 2 |
|---|---|---|---|---|

| | | made with vegetable fat | made with vegetable fat |
|---|---|---|---|
| % | Water | 2 | 2 |
| grams { | Calories | 515 | 10 |
| | Protein | 5 | Trace |
| | Fat | 33 | 1 |
| | Carbohydrate | 52 | 1 |
| | Calcium | 21 | Trace |
| milligrams { | Phosphorus | 397 | 8 |
| | Iron | .1 | Trace |
| | Sodium | N.A. | N.A. |
| | Potassium | 763 | 16 |
| | Thiamin | 0 | 0 |
| | Riboflavin | *.16 | *Trace |
| | Niacin | 0 | 0 |
| | Ascorbic Acid | 0 | 0 |
| unit | Vitamin A | *190 | *Trace |

*Vitamin A value is largely from beta-carotene used for coloring. Riboflavin value for powdered creams applies to products with added riboflavin.

# POWDERED
# WHIPPED TOPPING
## IMITATION

| grams | Weight: | 80 | 4 |
|---|---|---|---|

| | | made with whole milk | made with whole milk |
|---|---|---|---|
| % | Water | 67 | 67 |
| | Calories | 150 | 10 |
| grams | Protein | 3 | Trace |
| | Fat | 10 | Trace |
| | Carbohydrate | 13 | 1 |
| | Calcium | 72 | 4 |
| milligrams | Phosphorus | 69 | 3 |
| | Iron | Trace | Trace |
| | Sodium | N.A. | N.A. |
| | Potassium | 121 | 6 |
| | Thiamin | .02 | Trace |
| | Riboflavin | .09 | Trace |
| | Niacin | Trace | Trace |
| | Ascorbic Acid | 1 | Trace |
| unit | Vitamin A | *290 | *10 |

*Vitamin A value is largely from beta-carotene used for coloring.

# PRESSURIZED
# WHIPPED TOPPING
## IMITATION

grams | **Weight:** | 70 | 4

Pressurized | Pressurized

| | | | |
|---|---|---:|---:|
| % | **Water** | 60 | 60 |
| | **Calories** | 185 | 10 |
| grams | **Protein** | 1 | Trace |
| | **Fat** | 16 | 1 |
| | **Carbohydrate** | 11 | 1 |
| | **Calcium** | 4 | Trace |
| | **Phosphorus** | 13 | 1 |
| | **Iron** | Trace | Trace |
| | **Sodium** | N.A. | N.A. |
| milligrams | **Potassium** | 13 | 1 |
| | **Thiamin** | 0 | 0 |
| | **Riboflavin** | 0 | 0 |
| | **Niacin** | 0 | 0 |
| | **Ascorbic Acid** | 0 | 0 |
| unit | **Vitamin A** | *330 | *20 |

*Vitamin A value is largely from beta-carotene used for coloring.

# IMITATION
# SOUR DRESSING

| grams | Weight: | 235 | 12 |
|---|---|---|---|

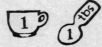

| | | made with nonfat dry milk | made with nonfat dry milk |
|---|---|---|---|
| % | Water | 75 | 75 |
| | Calories | 415 | 20 |
| grams | Protein | 8 | Trace |
| | Fat | 39 | 2 |
| | Carbohydrate | 11 | 1 |
| | Calcium | 266 | 14 |
| milligrams | Phosphorus | 205 | 10 |
| | Iron | .1 | Trace |
| | Sodium | N.A. | N.A. |
| | Potassium | 380 | 19 |
| | Thiamin | .09 | .01 |
| | Riboflavin | .38 | .02 |
| | Niacin | .2 | Trace |
| | Ascorbic Acid | 2 | Trace |
| unit | Vitamin A | *20 | *Trace |

*Vitamin A value is largely from beta-carotene used for coloring.

42

# CHEESE AND CHEESE PRODUCTS

BLUE CHEESE
BRICK CHEESE
CAMEMBERT CHEESE
CHEDDAR CHEESE
CHEESE SOUFFLE
COTTAGE CHEESE, CREAMED
COTTAGE CHEESE, LOWFAT
COTTAGE CHEESE, UNCREAMED
CREAM CHEESE
LIMBURGER CHEESE
MOZZARELLA CHEESE
PARMESAN CHEESE
PASTURIZED PROCESS CHEESE
PASTURIZED PROCESS CHEESE FOOD
PASTURIZED PROCESS CHEESE SPREAD
PROVOLONE CHEESE
RICOTTA CHEESE
ROMANO CHEESE
SWISS CHEESE

# BLUE CHEESE

grams     **Weight:**        28

cut pieces

| | | |
|---|---|---|
| % | **Water** | 42 |
| | **Calories** | 100 |
| grams | **Protein** | 6 |
| | **Fat** | 8 |
| | **Carbohydrate** | 1 |
| | **Calcium** | 150 |
| | **Phosphorus** | 110 |
| | **Iron** | .1 |
| | **Sodium** | N.A. |
| milligrams | **Potassium** | 73 |
| | **Thiamin** | .01 |
| | **Riboflavin** | .11 |
| | **Niacin** | .3 |
| | **Ascorbic Acid** | 0 |
| unit | **Vitamin A** | 200 |

44

# BRICK CHEESE

| grams | **Weight:** | 28 | 45 |
|---|---|---|---|

1 prepackaged slice (wt., 1 ½ oz.) 7 ⅛" x 3 ¾" x 3/32"

| | | 28 | 45 |
|---|---|---|---|
| % | **Water** | 41 | 41 |
| grams | **Calories** | 105 | 167 |
| | **Protein** | 6.3 | 10 |
| | **Fat** | 8.6 | 13.7 |
| | **Carbohydrate** | .5 | .9 |
| | **Calcium** | 207 | 329 |
| milligrams | **Phosphorus** | 129 | 205 |
| | **Iron** | .3 | .4 |
| | **Sodium** | N.A. | N.A. |
| | **Potassium** | N.A. | N.A. |
| | **Thiamin** | N.A. | N.A. |
| | **Riboflavin** | .13 | .20 |
| | **Niacin** | Trace | Trace |
| | **Ascorbic Acid** | 0 | 0 |
| unit | **Vitamin A** | 350 | 560 |

45

# CAMEMBERT CHEESE

grams | **Weight:** | 38

1 wedge

| | | |
|---|---|---|
| % | **Water** | 52 |
| | **Calories** | 115 |
| | **Protein** | 8 |
| grams | **Fat** | 9 |
| | **Carbohydrate** | Trace |
| | **Calcium** | 147 |
| | **Phosphorus** | 132 |
| | **Iron** | .1 |
| | **Sodium** | N.A. |
| | **Potassium** | 71 |
| milligrams | **Thiamin** | .01 |
| | **Riboflavin** | .19 |
| | **Niacin** | .2 |
| | **Ascorbic Acid** | 0 |
| unit | **Vitamin A** | 350 |

# CHEDDAR CHEESE

| grams | **Weight:** | 28 | 113 |
|---|---|---|---|

cut pieces | shredded

| | | 28 | 113 |
|---|---|---|---|
| % | **Water** | 37 | 37 |
| grams | **Calories** | 115 | 455 |
| | **Protein** | 7 | 28 |
| | **Fat** | 9 | 37 |
| | **Carbohydrate** | Trace | 1 |
| milligrams | **Calcium** | 204 | 815 |
| | **Phosphorus** | 145 | 579 |
| | **Iron** | .2 | .8 |
| | **Sodium** | 198 | 791 |
| | **Potassium** | 28 | 111 |
| | **Thiamin** | .01 | .03 |
| | **Riboflavin** | .11 | .42 |
| | **Niacin** | Trace | .1 |
| | **Ascorbic Acid** | 0 | 0 |
| unit | **Vitamin A** | 300 | 1,200 |

# CHEESE SOUFFLE

grams    **Weight:**      110

homemade,
baked in 8" sq.
pan or 7" dia.
casserole

| | | |
|---|---|---|
| % | **Water** | 65 |
| | **Calories** | 959 |
| grams { | **Protein** | 43.6 |
| | **Fat** | 75.2 |
| | **Carbohydrate** | 27.3 |
| | **Calcium** | 884 |
| | **Phosphorus** | 858 |
| | **Iron** | 4.4 |
| | **Sodium** | 1,602 |
| milligrams | **Potassium** | 532 |
| | **Thiamin** | .22 |
| | **Riboflavin** | 1.06 |
| | **Niacin** | .9 |
| | **Ascorbic Acid** | Trace |
| unit | **Vitamin A** | 3,520 |

# CREAMED
# COTTAGE CHEESE
## (4% FAT)

| | small curd* | large curd* |
|---|---|---|
| grams **Weight:** | 210 | 225 |

| | | small curd* | large curd* |
|---|---|---|---|
| % | **Water** | 79 | 79 |
| grams | **Calories** | 220 | 235 |
| | **Protein** | 26 | 28 |
| | **Fat** | 9 | 10 |
| | **Carbohydrate** | 6 | 6 |
| | **Calcium** | 126 | 135 |
| milligrams | **Phosphorus** | 277 | 297 |
| | **Iron** | .3 | .3 |
| | **Sodium** | 481 | 515 |
| | **Potassium** | 177 | 190 |
| | **Thiamin** | .04 | .05 |
| | **Riboflavin** | .34 | .37 |
| | **Niacin** | .3 | .3 |
| | **Ascorbic Acid** | Trace | Trace |
| unit | **Vitamin A** | 340 | 370 |

*Curd not pressed down.

49

# LOWFAT
# COTTAGE CHEESE
## (2% AND 1% FAT)

| grams | Weight: | 226 | 226 |
|---|---|---|---|

|  |  | 2% fat* | 1% fat |
|---|---|---|---|
| % | Water | 79 | 82 |
|  | Calories | 205 | 165 |
| grams { | Protein | 31 | 28 |
|  | Fat | 4 | 2 |
|  | Carbohydrate | 8 | 6 |
|  | Calcium | 155 | 138 |
| milligrams { | Phosphorus | 340 | 302 |
|  | Iron | .4 | .3 |
|  | Sodium | 561 | N.A. |
|  | Potassium | 217 | 193 |
|  | Thiamin | .05 | .05 |
|  | Riboflavin | .42 | .37 |
|  | Niacin | .3 | .3 |
|  | Ascorbic Acid | Trace | Trace |
| unit | Vitamin A | 160 | 80 |

*Curd not pressed down.

# UNCREAMED
# COTTAGE CHEESE
## (½% FAT)

grams **Weight:** 145

| | | |
|---|---|---|
| % | **Water** | 80 |
| | **Calories** | 125 |
| grams | **Protein** | 25 |
| | **Fat** | 1 |
| | **Carbohydrate** | 3 |
| | **Calcium** | 46 |
| milligrams | **Phosphorus** | 151 |
| | **Iron** | .3 |
| | **Sodium** | 421 |
| | **Potassium** | 47 |
| | **Thiamin** | .04 |
| | **Riboflavin** | .21 |
| | **Niacin** | .2 |
| | **Ascorbic Acid** | 0 |
| unit | **Vitamin A** | 40 |

# CREAM CHEESE

grams | **Weight:** | 28

| % | **Water** | 54 |
|---|---|---|
| | **Calories** | 100 |
| grams | **Protein** | 2 |
| | **Fat** | 10 |
| | **Carbohydrate** | 1 |
| | **Calcium** | 23 |
| | **Phosphorus** | 30 |
| | **Iron** | .3 |
| milligrams | **Sodium** | 71 |
| | **Potassium** | 34 |
| | **Thiamin** | Trace |
| | **Riboflavin** | .06 |
| | **Niacin** | Trace |
| | **Ascorbic Acid** | 0 |
| unit | **Vitamin A** | 400 |

# LIMBURGER CHEESE

grams | **Weight:** | 28

| % | **Water** | 45 |
|---|---|---|
| | **Calories** | 98 |
| grams | **Protein** | 6.0 |
| | **Fat** | 7.9 |
| | **Carbohydrate** | .6 |
| | **Calcium** | 167 |
| | **Phosphorus** | 111 |
| | **Iron** | .2 |
| | **Sodium** | N.A. |
| milligrams | **Potassium** | N.A. |
| | **Thiamin** | .02 |
| | **Riboflavin** | .14 |
| | **Niacin** | .1 |
| | **Ascorbic Acid** | 0 |
| unit | **Vitamin A** | 320 |

53

# MOZZARELLA CHEESE

| | grams | 28 | 28 |
|---|---|---|---|
| **Weight:** | | made with whole milk | made with part skim milk |
| Water | % | 48 | 48 |
| Calories | | 90 | 80 |
| Protein | grams | 6 | 8 |
| Fat | | 7 | 5 |
| Carbohydrate | | 1 | 1 |
| Calcium | | 163 | 207 |
| Phosphorus | milligrams | 117 | 149 |
| Iron | | .1 | .1 |
| Sodium | | N.A. | N.A. |
| Potassium | | 21 | 27 |
| Thiamin | | Trace | .01 |
| Riboflavin | | .08 | .10 |
| Niacin | | Trace | Trace |
| Ascorbic Acid | | 0 | 0 |
| Vitamin A | unit | 260 | 180 |

# PARMESAN CHEESE

| grams | **Weight:** | 100 | 5 | 28 |
|---|---|---|---|---|

not pressed
down

| | | | | |
|---|---|---|---|---|
| % | **Water** | 18 | 18 | 18 |
| | **Calories** | 455 | 25 | 130 |
| grams | **Protein** | 42 | 2 | 12 |
| | **Fat** | 30 | 2 | 9 |
| | **Carbohydrate** | 4 | Trace | 1 |
| | **Calcium** | 1,376 | 69 | 390 |
| | **Phosphorus** | 807 | 40 | 229 |
| | **Iron** | 1.0 | Trace | .3 |
| milligrams | **Sodium** | 870 | 44 | 247 |
| | **Potassium** | 107 | 5 | 30 |
| | **Thiamin** | .05 | Trace | .01 |
| | **Riboflavin** | .39 | .02 | .11 |
| | **Niacin** | .3 | Trace | .1 |
| | **Ascorbic Acid** | 0 | 0 | 0 |
| unit | **Vitamin A** | 700 | 40 | 200 |

# PASTEURIZED PROCESS
# CHEESE

| | grams | **Weight:** | 28 | 28 |
|---|---|---|---|---|

| | | American | Swiss |
|---|---|---|---|

| | | American | Swiss |
|---|---|---|---|
| % | **Water** | 39 | 42 |
| | **Calories** | 105 | 95 |
| grams | **Protein** | 6 | 7 |
| | **Fat** | 9 | 7 |
| | **Carbohydrate** | Trace | 1 |
| | **Calcium** | 174 | 219 |
| milligrams | **Phosphorus** | 211 | 216 |
| | **Iron** | .1 | .2 |
| | **Sodium** | 322 | 331 |
| | **Potassium** | 46 | 61 |
| | **Thiamin** | .01 | Trace |
| | **Riboflavin** | .10 | .08 |
| | **Niacin** | Trace | Trace |
| | **Ascorbic Acid** | 0 | 0 |
| unit | **Vitamin A** | 340 | 230 |

# PASTEURIZED PROCESS
# CHEESE FOOD

grams    **Weight:**     28

American

| | | |
|---|---|---|
| % | **Water** | 43 |
| | **Calories** | 95 |
| | **Protein** | 6 |
| grams | **Fat** | 7 |
| | **Carbohydrate** | 2 |
| | **Calcium** | 163 |
| | **Phosphorus** | 130 |
| | **Iron** | .2 |
| | **Sodium** | 128 |
| milligrams | **Potassium** | 79 |
| | **Thiamin** | .01 |
| | **Riboflavin** | .13 |
| | **Niacin** | Trace |
| | **Ascorbic Acid** | 0 |
| unit | **Vitamin A** | 260 |

# PASTEURIZED PROCESS
# CHEESE SPREAD

grams   **Weight:**        28

American

| | | |
|---|---|---|
| % | **Water** | 48 |
| | **Calories** | 82 |
| grams { | **Protein** | 5 |
| | **Fat** | 6 |
| | **Carbohydrate** | 2 |
| | **Calcium** | 159 |
| milligrams { | **Phosphorus** | 202 |
| | **Iron** | .1 |
| | **Sodium** | 461 |
| | **Potassium** | 69 |
| | **Thiamin** | .01 |
| | **Riboflavin** | .12 |
| | **Niacin** | Trace |
| | **Ascorbic Acid** | 0 |
| unit | **Vitamin A** | 220 |

# PROVOLONE CHEESE

| | Weight: | 28 |
|---|---|---|
| grams | | |

cut pieces

| | | |
|---|---|---|
| % | **Water** | 41 |
| | **Calories** | 100 |
| grams | **Protein** | 7 |
| | **Fat** | 8 |
| | **Carbohydrate** | 1 |
| | **Calcium** | 214 |
| milligrams | **Phosphorus** | 141 |
| | **Iron** | .1 |
| | **Sodium** | N.A. |
| | **Potassium** | 39 |
| | **Thiamin** | .01 |
| | **Riboflavin** | .09 |
| | **Niacin** | Trace |
| | **Ascorbic Acid** | 0 |
| unit | **Vitamin A** | 230 |

# RICOTTA CHEESE

| | | made from whole milk | made from part skim milk |
|---|---|---|---|
| grams | **Weight:** | 246 | 246 |

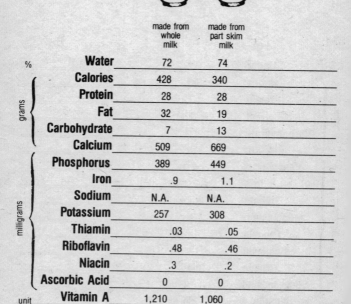

| | | made from whole milk | made from part skim milk |
|---|---|---|---|
| % | **Water** | 72 | 74 |
| grams | **Calories** | 428 | 340 |
| | **Protein** | 28 | 28 |
| | **Fat** | 32 | 19 |
| | **Carbohydrate** | 7 | 13 |
| | **Calcium** | 509 | 669 |
| milligrams | **Phosphorus** | 389 | 449 |
| | **Iron** | .9 | 1.1 |
| | **Sodium** | N.A. | N.A. |
| | **Potassium** | 257 | 308 |
| | **Thiamin** | .03 | .05 |
| | **Riboflavin** | .48 | .46 |
| | **Niacin** | .3 | .2 |
| | **Ascorbic Acid** | 0 | 0 |
| unit | **Vitamin A** | 1,210 | 1,060 |

# ROMANO CHEESE

grams **Weight:** 28

cut pieces

| | | |
|---|---|---|
| % | **Water** | 31 |
| | **Calories** | 110 |
| grams { | **Protein** | 9 |
| | **Fat** | 8 |
| | **Carbohydrate** | 1 |
| | **Calcium** | 302 |
| | **Phosphorus** | 215 |
| | **Iron** | — |
| milligrams { | **Sodium** | N.A. |
| | **Potassium** | N.A. |
| | **Thiamin** | N.A. |
| | **Riboflavin** | .11 |
| | **Niacin** | Trace |
| | **Ascorbic Acid** | 0 |
| unit | **Vitamin A** | 160 |

# SWISS CHEESE

grams **Weight:** 28

cut pieces

| | | |
|---|---|---|
| % | **Water** | 37 |
| | **Calories** | 105 |
| grams | **Protein** | 8 |
| | **Fat** | 8 |
| | **Carbohydrate** | 1 |
| | **Calcium** | 272 |
| | **Phosphorus** | 171 |
| | **Iron** | Trace |
| milligrams | **Sodium** | 201 |
| | **Potassium** | 31 |
| | **Thiamin** | .01 |
| | **Riboflavin** | .10 |
| | **Niacin** | Trace |
| | **Ascorbic Acid** | 0 |
| unit | **Vitamin A** | 240 |

# EGGS

RAW
FRIED IN BUTTER
HARD BOILED
POACHED
SCRAMBLED OR OMELETTE

# EGGS

| grams | Weight: | 50 | 33 | 17 |
|---|---|---|---|---|
| | | ①<br>*whole<br>without<br>shell | ①<br>white | ①<br>yolk |
| % | **Water** | 75 | 88 | 49 |
| | **Calories** | 80 | 15 | 65 |
| grams | **Protein** | 6 | 3 | 3 |
| | **Fat** | 6 | Trace | 6 |
| | **Carbohydrate** | 1 | Trace | Trace |
| | **Calcium** | 28 | 4 | 26 |
| milligrams | **Phosphorus** | 90 | 4 | 86 |
| | **Iron** | 1.0 | Trace | .9 |
| | **Sodium** | 61 | 48 | 9 |
| | **Potassium** | 65 | 45 | 15 |
| | **Thiamin** | .04 | Trace | .04 |
| | **Riboflavin** | .15 | .09 | .07 |
| | **Niacin** | Trace | Trace | Trace |
| | **Ascorbic Acid** | 0 | 0 | 0 |
| unit | **Vitamin A** | 260 | 0 | 310 |

*Large (24 oz. per dozen).

# EGGS

grams **Weight:** 46

*fried in
butter

| | |
|---|---|
| % | **Water** 72 |
| | **Calories** 85 |
| | **Protein** 5 |
| grams | **Fat** 6 |
| | **Carbohydrate** 1 |
| | **Calcium** 26 |
| | **Phosphorus** 80 |
| | **Iron** .9 |
| | **Sodium** 155 |
| milligrams | **Potassium** 58 |
| | **Thiamin** .03 |
| | **Riboflavin** .13 |
| | **Niacin** Trace |
| | **Ascorbic Acid** 0 |
| unit | **Vitamin A** 290 |

*Large (24 oz. per dozen).

# HARD BOILED
# EGGS
## SHELL REMOVED

grams   **Weight:**        50

*hard boiled,
shell removed

| % | | | |
|---|---|---|---|
| | **Water** | 75 | |
| | **Calories** | 80 | |
| | **Protein** | 6 | |
| | **Fat** | 6 | |
| | **Carbohydrate** | 1 | |
| | **Calcium** | 28 | |
| | **Phosphorus** | 90 | |
| | **Iron** | 1.0 | |
| | **Sodium** | 61 | |
| | **Potassium** | 65 | |
| | **Thiamin** | .04 | |
| | **Riboflavin** | .14 | |
| | **Niacin** | Trace | |
| | **Ascorbic Acid** | 0 | |
| | **Vitamin A** | 260 | |

grams { (Protein through Calcium)

milligrams { (Phosphorus through Ascorbic Acid)

unit

*Large (24 oz. per dozen).

# EGGS

| grams | **Weight:** | 50 |

*poached

| % | **Water** | 74 |
|---|---|---|
| | **Calories** | 80 |
| grams { | **Protein** | 6 |
| | **Fat** | 6 |
| | **Carbohydrate** | 1 |
| | **Calcium** | 28 |
| milligrams { | **Phosphorus** | 90 |
| | **Iron** | 1.0 |
| | **Sodium** | 136 |
| | **Potassium** | 65 |
| | **Thiamin** | .04 |
| | **Riboflavin** | .13 |
| | **Niacin** | Trace |
| | **Ascorbic Acid** | 0 |
| unit | **Vitamin A** | 260 |

*Large (24 oz. per dozen).

# EGGS

grams **Weight:** 64

*scrambled in
butter
(milk added)
(also omelette)

| | % | |
|---|---|---|
| **Water** | 76 | |

grams {

| | | |
|---|---|---|
| **Calories** | 95 | |
| **Protein** | 6 | |
| **Fat** | 7 | |
| **Carbohydrate** | 1 | |
| **Calcium** | 47 | |

milligrams {

| | | |
|---|---|---|
| **Phosphorus** | 97 | |
| **Iron** | .9 | |
| **Sodium** | 164 | |
| **Potassium** | 85 | |
| **Thiamin** | .04 | |
| **Riboflavin** | .16 | |
| **Niacin** | Trace | |
| **Ascorbic Acid** | 0 | |
| **Vitamin A** | 310 | |

unit

*Large (24 oz. per dozen).

# OILS AND FATS

BUTTER, REGULAR
BUTTER, WHIPPED
CORN OIL
LARD
MARGARINE, REGULAR
MARGARINE, SOFT
MARGARINE, WHIPPED
OLIVE OIL
PEANUT OIL
SAFFLOWER OIL
SOYBEAN-COTTONSEED OIL BLEND
SOYBEAN OIL
VEGETABLE SHORTENING

# BUTTER
## REGULAR

| grams | Weight: | 113 | 14 | 5 |
|---|---|---|---|---|
| | | ½ cup | about 1/8 stick | 1 in. sq. oz. ⅓ in. high; 90 per lb. |

| | | | | |
|---|---|---|---|---|
| % | Water | 16 | 16 | 16 |
| grams | Calories | 815 | 100 | 35 |
| | Protein | 1 | Trace | Trace |
| | Fat | 92 | 12 | 4 |
| | Carbohydrate | Trace | Trace | Trace |
| | Calcium | 27 | 3 | 1 |
| milligrams | Phosphorus | 26 | 3 | 1 |
| | Iron | .2 | Trace | Trace |
| | Sodium | 1,119 | 140 | 49 |
| | Potassium | 29 | 4 | 1 |
| | Thiamin | .01 | Trace | Trace |
| | Riboflavin | .04 | Trace | Trace |
| | Niacin | Trace | Trace | Trace |
| | Ascorbic Acid | 0 | 0 | 0 |
| unit | Vitamin A | *3,470 | *430 | *150 |

*Based on year-round average.

# BUTTER
## WHIPPED

| | | grams | | |
|---|---|:---:|:---:|:---:|
| grams | **Weight:** | 76 | 9 | 4 |
| | | ½ cup 1/8 stick | about ⅓ high, 120 per lb. | 1¼ in. sq. |

| | | % | | |
|---|---|:---:|:---:|:---:|
| % | **Water** | 16 | 16 | 16 |
| grams | **Calories** | 540 | 65 | 25 |
| | **Protein** | 1 | Trace | Trace |
| | **Fat** | 61 | 8 | 3 |
| | **Carbohydrate** | Trace | Trace | Trace |
| milligrams | **Calcium** | 18 | 2 | 1 |
| | **Phosphorus** | 17 | 2 | 1 |
| | **Iron** | .1 | Trace | Trace |
| | **Sodium** | 746 | 93 | 38 |
| | **Potassium** | 20 | 2 | 1 |
| | **Thiamin** | Trace | Trace | 0 |
| | **Riboflavin** | .03 | Trace | Trace |
| | **Niacin** | Trace | Trace | Trace |
| | **Ascorbic Acid** | 0 | 0 | 0 |
| unit | **Vitamin A** | *2,310 | *290 | *120 |

*Based on year-round average.

# CORN OIL

| | grams | | |
|---|---|---|---|
| **Weight:** | | 218 | 14 |

| | | | |
|---|---|---|---|
| % | **Water** | 0 | 0 |
| | **Calories** | 1,925 | 120 |
| | **Protein** | 0 | 0 |
| grams | **Fat** | 218 | 14 |
| | **Carbohydrate** | 0 | 0 |
| | **Calcium** | 0 | 0 |
| | **Phosphorus** | 0 | 0 |
| | **Iron** | 0 | 0 |
| | **Sodium** | 0 | 0 |
| milligrams | **Potassium** | 0 | 0 |
| | **Thiamin** | 0 | 0 |
| | **Riboflavin** | 0 | 0 |
| | **Niacin** | 0 | 0 |
| | **Ascorbic Acid** | 0 | 0 |
| unit | **Vitamin A** | N.A. | N.A. |

# LARD

grams **Weight:** 205 13

| | | | |
|---|---|---|---|
| % | Water | 0 | 0 |
| | Calories | 1,850 | 115 |
| grams | Protein | 0 | 0 |
| | Fat | 205 | 13 |
| | Carbohydrate | 0 | 0 |
| | Calcium | 0 | 0 |
| milligrams | Phosphorus | 0 | 0 |
| | Iron | 0 | 0 |
| | Sodium | 0 | 0 |
| | Potassium | 0 | 0 |
| | Thiamin | 0 | 0 |
| | Riboflavin | 0 | 0 |
| | Niacin | 0 | 0 |
| | Ascorbic Acid | 0 | 0 |
| unit | Vitamin A | 0 | 0 |

# MARGARINE
## REGULAR

| | Weight: | 113 | 14 | 5 |
|---|---|---|---|---|
grams

| | | ½ cup | about 1/8 stick | 1 in. sq., ⅓ in. high, 90 per lb. |
|---|---|---|---|---|

| | | | | |
|---|---|---|---|---|
| % | Water | 16 | 16 | 16 |
| | Calories | 815 | 100 | 35 |
| grams | Protein | 1 | Trace | Trace |
| | Fat | 92 | 12 | 4 |
| | Carbohydrate | Trace | Trace | Trace |
| | Calcium | 27 | 3 | 1 |
| milligrams | Phosphorus | 26 | 3 | 1 |
| | Iron | .2 | Trace | Trace |
| | Sodium | 1,119 | 140 | 49 |
| | Potassium | 29 | 4 | 1 |
| | Thiamin | .01 | Trace | Trace |
| | Riboflavin | .04 | Trace | Trace |
| | Niacin | Trace | Trace | Trace |
| | Ascorbic Acid | 0 | 0 | 0 |
| unit | Vitamin A | *3,750 | *470 | *170 |

*Based on average Vitamin A content of fortified margarine. Federal specifications for fortified margarine require a minimum of 15,000 International Units (I.U.) of Vitamin A per pound.

# MARGARINE
## SOFT

grams **Weight:** 227      14

1 container

| | | |
|---|---|---|
| **Water** % | 16 | 16 |
| **Calories** | 1,635 | 100 |
| **Protein** | 1 | Trace |
| **Fat** | 184 | 12 |
| **Carbohydrate** | Trace | Trace |
| **Calcium** | 53 | 3 |
| **Phosphorus** | 52 | 3 |
| **Iron** | .4 | Trace |
| **Sodium** | 2,238 | 140 |
| **Potassium** | 59 | 4 |
| **Thiamin** | .01 | Trace |
| **Riboflavin** | .08 | Trace |
| **Niacin** | .1 | Trace |
| **Ascorbic Acid** | 0 | 0 |
| **Vitamin A** unit | *7,500 | *470 |

*Based on average Vitamin A content of fortified margarine. Federal specifications for fortified margarine require a minimum of 15,000 International Units (I.U.) of Vitamin A per pound.

# MARGARINE
## WHIPPED

| | Weight: | 76 | 9 |
|---|---|---|---|
| grams | | | |

| | | ½ cup | about 1/8 stick |
|---|---|---|---|
| % | Water | 16 | 16 |
| | Calories | 545 | 70 |
| grams | Protein | Trace | Trace |
| | Fat | 61 | 8 |
| | Carbohydrate | Trace | Trace |
| | Calcium | 18 | 2 |
| milligrams | Phosphorus | 17 | 2 |
| | Iron | .1 | Trace |
| | Sodium | 746 | 93 |
| | Potassium | 20 | 2 |
| | Thiamin | Trace | Trace |
| | Riboflavin | 03 | Trace |
| | Niacin | Trace | Trace |
| | Ascorbic Acid | 0 | 0 |
| unit | Vitamin A | *2,500 | *310 |

*Based on average Vitamin A content of fortified margarine. Federal specifications for fortified margarine require a minimum of 15,000 I.U. of Vitamin A per pound.

# OLIVE OIL

grams  **Weight:**  216  14

| | | |
|---|---|---|
| % | **Water** | 0 | 0 |
| | **Calories** | 1,910 | 120 |
| grams | **Protein** | 0 | 0 |
| | **Fat** | 216 | 14 |
| | **Carbohydrate** | 0 | 0 |
| | **Calcium** | 0 | 0 |
| milligrams | **Phosphorus** | 0 | 0 |
| | **Iron** | 0 | 0 |
| | **Sodium** | 0 | 0 |
| | **Potassium** | 0 | 0 |
| | **Thiamin** | 0 | 0 |
| | **Riboflavin** | 0 | 0 |
| | **Niacin** | 0 | 0 |
| | **Ascorbic Acid** | 0 | 0 |
| unit | **Vitamin A** | N.A. | N.A. |

# PEANUT OIL

| | grams Weight: | 216 | 14 |
|---|---|---|---|

| | | | |
|---|---|---|---|
| % | **Water** | 0 | 0 |
| | **Calories** | 1,910 | 120 |
| grams | **Protein** | 0 | 0 |
| | **Fat** | 216 | 14 |
| | **Carbohydrate** | 0 | 0 |
| | **Calcium** | 0 | 0 |
| milligrams | **Phosphorus** | 0 | 0 |
| | **Iron** | 0 | 0 |
| | **Sodium** | 0 | 0 |
| | **Potassium** | 0 | 0 |
| | **Thiamin** | 0 | 0 |
| | **Riboflavin** | 0 | 0 |
| | **Niacin** | 0 | 0 |
| | **Ascorbic Acid** | 0 | 0 |
| unit | **Vitamin A** | N.A. | N.A. |

# SAFFLOWER OIL

| | grams | Weight: | 218 | 14 |
|---|---|---|---|---|

| | | 1 cup | 1 tbs |
|---|---|---|---|
| % | Water | 0 | 0 |
| | Calories | 1,925 | 120 |
| grams | Protein | 0 | 0 |
| | Fat | 218 | 14 |
| | Carbohydrate | 0 | 0 |
| | Calcium | 0 | 0 |
| milligrams | Phosphorus | 0 | 0 |
| | Iron | 0 | 0 |
| | Sodium | 0 | 0 |
| | Potassium | 0 | 0 |
| | Thiamin | 0 | 0 |
| | Riboflavin | 0 | 0 |
| | Niacin | 0 | 0 |
| | Ascorbic Acid | 0 | 0 |
| unit | Vitamin A | N.A. | N.A. |

# SOYBEAN
## COTTONSEED OIL BLEND

| grams | Weight: | 218 | 14 |
|---|---|---|---|

| | | | |
|---|---|---|---|
| % | **Water** | 0 | 0 |
| | **Calories** | 1,925 | 120 |
| grams | **Protein** | 0 | 0 |
| | **Fat** | 218 | 14 |
| | **Carbohydrate** | 0 | 0 |
| | **Calcium** | 0 | 0 |
| milligrams | **Phosphorus** | 0 | 0 |
| | **Iron** | 0 | 0 |
| | **Sodium** | 0 | 0 |
| | **Potassium** | 0 | 0 |
| | **Thiamin** | 0 | 0 |
| | **Riboflavin** | 0 | 0 |
| | **Niacin** | 0 | 0 |
| | **Ascorbic Acid** | 0 | 0 |
| unit | **Vitamin A** | N.A. | N.A. |

# SOYBEAN OIL

| | Weight: | 218 | 14 |
|---|---|---|---|

grams

| | | Partially hardened, hydrogenated | Partially hardened hydrogenated |
|---|---|---|---|
| % | Water | 0 | 0 |
| | Calories | 1,925 | 120 |
| | Protein | 0 | 0 |
| grams | Fat | 218 | 218 |
| | Carbohydrate | 0 | 0 |
| | Calcium | 0 | 0 |
| | Phosphorus | 0 | 0 |
| | Iron | 0 | 0 |
| | Sodium | 0 | 0 |
| milligrams | Potassium | 0 | 0 |
| | Thiamin | 0 | 0 |
| | Riboflavin | 0 | 0 |
| | Niacin | 0 | 0 |
| | Ascorbic Acid | 0 | 0 |
| unit | Vitamin A | N.A. | N.A. |

# VEGETABLE
# SHORTENING

grams **Weight:** 200 13

| | | | |
|---|---|---|---|
| % | **Water** | 0 | 0 |
| | **Calories** | 1,770 | 110 |
| grams | **Protein** | 0 | 0 |
| | **Fat** | 200 | 13 |
| | **Carbohydrate** | 0 | 0 |
| | **Calcium** | 0 | 0 |
| milligrams | **Phosphorus** | 0 | 0 |
| | **Iron** | 0 | 0 |
| | **Sodium** | 0 | 0 |
| | **Potassium** | 0 | 0 |
| | **Thiamin** | 0 | 0 |
| | **Riboflavin** | 0 | 0 |
| | **Niacin** | 0 | 0 |
| | **Ascorbic Acid** | 0 | 0 |
| unit | **Vitamin A** | N.A. | N.A. |

# FISH AND SHELLFISH

ANCHOVIES
BASS
BLUEFISH
CAVIAR
CLAMS
CRAB
FISH STICKS
HADDOCK
OCEAN PERCH
OYSTERS
SALMON
SARDINES
SCALLOPS
SHAD
SHRIMP
TUNA
TUNA SALAD

# ANCHOVIES

grams **Weight:** 20

flat or rolled,
canned

| | | |
|---|---|---|
| % | **Water** | 58.6 |
| | **Calories** | 35 |
| grams { | **Protein** | 3.8 |
| | **Fat** | 2.1 |
| | **Carbohydrate** | .1 |
| | **Calcium** | 34 |
| | **Phosphorus** | 42 |
| | **Iron** | 4 |
| milligrams { | **Sodium** | N.A. |
| | **Potassium** | N.A. |
| | **Thiamin** | N.A. |
| | **Riboflavin** | N.A. |
| | **Niacin** | N.A. |
| | **Ascorbic Acid** | 6 |
| unit | **Vitamin A** | N.A. |

# STRIPED BASS

| | | |
|---|---|---|
| grams | **Weight:** | 200 |

oven fried

| | | |
|---|---|---|
| % | **Water** | 60.8 |
| | **Calories** | 392 |
| grams { | **Protein** | 43 |
| | **Fat** | 17 |
| | **Carbohydrate** | 13.4 |
| | **Calcium** | 2 |
| milligrams { | **Phosphorus** | N.A. |
| | **Iron** | 4 |
| | **Sodium** | N.A. |
| | **Potassium** | N.A. |
| | **Thiamin** | N.A. |
| | **Riboflavin** | N.A. |
| | **Niacin** | 2 |
| | **Ascorbic Acid** | N.A. |
| unit | **Vitamin A** | N.A. |

# BLUEFISH

| | | |
|---|---|---|
| grams | **Weight:** | 85 |

OZ. 3

Baked with
butter or
margarine

| | | |
|---|---|---|
| % | **Water** | 68 |
| | **Calories** | 135 |
| | **Protein** | 22 |
| grams | **Fat** | 4 |
| | **Carbohydrate** | 0 |
| | **Calcium** | 25 |
| | **Phosphorus** | 244 |
| | **Iron** | 0.6 |
| | **Sodium** | *87 |
| milligrams | **Potassium** | N.A. |
| | **Thiamin** | .09 |
| | **Riboflavin** | .08 |
| | **Niacin** | 1.6 |
| | **Ascorbic Acid** | N.A. |
| unit | **Vitamin A** | 40 |

*Value for product without salt added.

86

# CAVIAR
## STURGEON

| | | granular | pressed |
|---|---|---|---|
| grams | **Weight:** | 16 | 17 |

| % | **Water** | 46 | 46 |
|---|---|---|---|
| | **Calories** | 42 | 54 |
| grams | **Protein** | 4.3 | 5.8 |
| | **Fat** | 2.4 | 2.8 |
| | **Carbohydrate** | .5 | .8 |
| | **Calcium** | 44 | N.A. |
| milligrams | **Phosphorus** | 57 | N.A. |
| | **Iron** | 1.9 | N.A. |
| | **Sodium** | 352 | N.A. |
| | **Potassium** | 29 | N.A. |
| | **Thiamin** | N.A. | N.A. |
| | **Riboflavin** | N.A. | N.A. |
| | **Niacin** | N.A. | N.A. |
| | **Ascorbic Acid** | N.A. | N.A. |
| unit | **Vitamin A** | N.A. | N.A. |

# CLAMS

| | Weight: | 85 | 85 |
|---|---|---|---|
| grams | | | |

|  | | OZ. 3 | OZ. 3 |
|---|---|---|---|
| | | raw meat only | canned solids and liquid |

| | | | |
|---|---|---|---|
| % | **Water** | 82 | 86 |
| | **Calories** | 65 | 45 |
| grams | **Protein** | 11 | 7 |
| | **Fat** | 1 | 1 |
| | **Carbohydrate** | 2 | 2 |
| | **Calcium** | 59 | 47 |
| milligrams | **Phosphorus** | 138 | 116 |
| | **Iron** | 5.2 | 3.5 |
| | **Sodium** | 25 | N.A. |
| | **Potassium** | 154 | 119 |
| | **Thiamin** | .08 | .01 |
| | **Riboflavin** | .15 | .09 |
| | **Niacin** | 1.1 | .9 |
| | **Ascorbic Acid** | 8 | N.A. |
| unit | **Vitamin A** | 90 | N.A. |

# CRAB MEAT

| | Weight: | 135 | 115 |
|---|---|---|---|
| grams | | | |

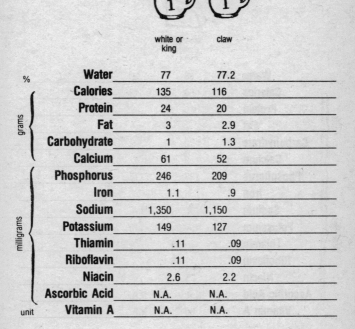

| | | white or king | claw |
|---|---|---|---|
| % | Water | 77 | 77.2 |
| | Calories | 135 | 116 |
| grams | Protein | 24 | 20 |
| | Fat | 3 | 2.9 |
| | Carbohydrate | 1 | 1.3 |
| | Calcium | 61 | 52 |
| milligrams | Phosphorus | 246 | 209 |
| | Iron | 1.1 | .9 |
| | Sodium | 1,350 | 1,150 |
| | Potassium | 149 | 127 |
| | Thiamin | .11 | .09 |
| | Riboflavin | .11 | .09 |
| | Niacin | 2.6 | 2.2 |
| | Ascorbic Acid | N.A. | N.A. |
| unit | Vitamin A | N.A. | N.A. |

# FISH STICKS

| grams | **Weight:** | 28 |

breaded,
cooked,
frozen**

| | | breaded, cooked, frozen** |
|---|---|---|
| % | **Water** | 66 |
| | **Calories** | 50 |
| grams | **Protein** | 5 |
| | **Fat** | 3 |
| | **Carbohydrate** | 2 |
| | **Calcium** | 3 |
| | **Phosphorus** | 47 |
| | **Iron** | .1 |
| | **Sodium** | N.A. |
| milligrams | **Potassium** | N.A. |
| | **Thiamin** | .01 |
| | **Riboflavin** | .02 |
| | **Niacin** | 5. |
| | **Ascorbic Acid** | N.A. |
| unit | **Vitamin A** | 0 |

**Stick, 4 by 1 by ½ in.

# HADDOCK

| | | |
|---|---|---|
| grams | **Weight:** | 85 |

breaded,
fried*

| % | **Water** | 66 |
|---|---|---|
| | **Calories** | 140 |
| grams | **Protein** | 17 |
| | **Fat** | 5 |
| | **Carbohydrate** | 5 |
| | **Calcium** | 34 |
| | **Phosphorus** | 210 |
| | **Iron** | 1.0 |
| milligrams | **Sodium** | 150** |
| | **Potassium** | 296 |
| | **Thiamin** | .03 |
| | **Riboflavin** | .06 |
| | **Niacin** | 2.7 |
| | **Ascorbic Acid** | 2 |
| unit | **Vitamin A** | N.A. |

*Dipped in egg, milk or water and bread crumbs; fried in vegetable shortening.
**Value for product without salt added.

# ATLANTIC
# OCEAN PERCH
## (RED FISH)

| | grams | **Weight:** | 85 |
|---|---|---|---|

breaded,
fried*

| | | |
|---|---|---|
| % | **Water** | 59 |
| | **Calories** | 195 |
| grams | **Protein** | 16 |
| | **Fat** | 11 |
| | **Carbohydrate** | 6 |
| | **Calcium** | 28 |
| | **Phosphorus** | 192 |
| | **Iron** | 1.1 |
| milligrams | **Sodium** | 130 |
| | **Potassium** | 242 |
| | **Thiamin** | .10 |
| | **Riboflavin** | .10 |
| | **Niacin** | 1.6 |
| | **Ascorbic Acid** | N.A. |
| unit | **Vitamin A** | N.A. |

*Dipped in egg, milk or water, and bread crumbs; fried in vegetable shortening.

# OYSTERS

| | Weight: | 240 | 240 |
|---|---|---|---|
| grams | |  |  |
| | | Eastern* | Pacific and Western (Olympic)** |
| % | Water | 85 | 79.1 |
| | **Calories** | 160 | 218 |
| grams | **Protein** | 20 | 25.4 |
| | **Fat** | 4 | 5.3 |
| | **Carbohydrate** | 8 | 15.4 |
| | **Calcium** | 226 | 204 |
| milligrams | **Phosphorus** | 343 | 367 |
| | **Iron** | 13.2 | 17.3 |
| | **Sodium** | 175 | N.A. |
| | **Potassium** | 290 | N.A. |
| | **Thiamin** | .34 | .29 |
| | **Riboflavin** | .43 | N.A. |
| | **Niacin** | 6.0 | 3.1 |
| | **Ascorbic Acid** | N.A. | 72 |
| unit | **Vitamin A** | 740 | N.A. |

*13-19 medium selects.
**4-6 medium, 6-9 small.

# SALMON
## PINK

grams  **Weight:**  85

| | | |
|---|---|---|
| % | **Water** | 71 |
| | **Calories** | 120 |
| grams | **Protein** | 17 |
| | **Fat** | 5 |
| | **Carbohydrate** | 0 |
| | **Calcium** | *167 |
| milligrams | **Phosphorus** | 243 |
| | **Iron** | .7 |
| | **Sodium** | 580 |
| | **Potassium** | 307 |
| | **Thiamin** | .03 |
| | **Riboflavin** | .16 |
| | **Niacin** | 6.8 |
| | **Ascorbic Acid** | N.A. |
| unit | **Vitamin A** | 60 |

*If bones are discarded value for calcium will be greatly reduced. Includes solids and liquids.

# ATLANTIC
# SARDINES

| | | |
|---|---|---|
| grams | **Weight:** | 85 |

canned in oil
drained solids

| | | |
|---|---|---|
| % | **Water** | 62 |
| | **Calories** | 175 |
| grams | **Protein** | 20 |
| | **Fat** | 9 |
| | **Carbohydrate** | 0 |
| | **Calcium** | 372 |
| | **Phosphorus** | 424 |
| | **Iron** | 25 |
| | **Sodium** | 699 |
| milligrams | **Potassium** | 502 |
| | **Thiamin** | .02 |
| | **Riboflavin** | .17 |
| | **Niacin** | 4.6 |
| | **Ascorbic Acid** | N.A. |
| unit | **Vitamin A** | 190 |

# SCALLOPS

| | | |
|---|---|---|
| grams | **Weight:** | 90 |

frozen*,
breaded,
fried,
reheated

| % | **Water** | 60 |
|---|---|---|
| | **Calories** | 175 |
| | **Protein** | 16 |
| | **Fat** | 8 |
| | **Carbohydrate** | 9 |
| | **Calcium** | N.A. |
| | **Phosphorus** | N.A. |
| | **Iron** | N.A. |
| | **Sodium** | N.A. |
| | **Potassium** | N.A. |
| | **Thiamin** | N.A. |
| | **Riboflavin** | N.A. |
| | **Niacin** | N.A. |
| | **Ascorbic Acid** | N.A. |
| | **Vitamin A** | N.A. |

grams { (Protein, Fat, Carbohydrate)

milligrams { (Calcium through Ascorbic Acid)

unit (Vitamin A)

*6 scallops.

# SHAD

| | | baked in butter or margarine, bacon |
|---|---|---|
| grams | **Weight:** | 85 |
| % | **Water** | 64 |
| | **Calories** | 170 |
| grams | **Protein** | 20 |
| | **Fat** | 10 |
| | **Carbohydrate** | 0 |
| | **Calcium** | 20 |
| milligrams | **Phosphorus** | 266 |
| | **Iron** | .5 |
| | **Sodium** | *66 |
| | **Potassium** | 320 |
| | **Thiamin** | .11 |
| | **Riboflavin** | .22 |
| | **Niacin** | 73 |
| | **Ascorbic Acid** | N.A. |
| unit | **Vitamin A** | 30 |

*Value for product without salt added.

# SHRIMP

| | Weight: | 85 | 85 |
|---|---|---|---|
| grams | | shrimp canned meat | shrimp french fried* |
| % | Water | 70 | 57 |
| | Calories | 100 | 190 |
| grams | Protein | 21 | 17 |
| | Fat | 1 | 9 |
| | Carbohydrate | 1 | 9 |
| | Calcium | 98 | 61 |
| milligrams | Phosphorus | 224 | 162 |
| | Iron | 2.6 | 1.7 |
| | Sodium | N.A. | **159 |
| | Potassium | 104 | 195 |
| | Thiamin | .01 | .03 |
| | Riboflavin | .03 | .07 |
| | Niacin | 1.5 | 2.3 |
| | Ascorbic Acid | N.A. | N.A. |
| unit | Vitamin A | 50 | N.A. |

*Dipped in egg, bread crumbs, and flour or butter.
**Value for product without salt added.

98

# TUNA

| grams | **Weight:** | 85 |

canned in oil,
drained solids

| | | |
|---|---|---|
| % | **Water** | 61 |
| | **Calories** | 170 |
| grams | **Protein** | 24 |
| | **Fat** | 7 |
| | **Carbohydrate** | 0 |
| | **Calcium** | 7 |
| | **Phosphorus** | 199 |
| | **Iron** | 1.6 |
| | **Sodium** | 465 |
| milligrams | **Potassium** | N.A. |
| | **Thiamin** | .04 |
| | **Riboflavin** | .10 |
| | **Niacin** | 10.1 |
| | **Ascorbic Acid** | N.A. |
| unit | **Vitamin A** | 70 |

# TUNA SALAD

grams **Weight:** 205

*

| | | |
|---|---|---|
| % | **Water** | 70 |
| grams { | **Calories** | 350 |
| | **Protein** | 30 |
| | **Fat** | 22 |
| | **Carbohydrate** | 7 |
| | **Calcium** | 41 |
| milligrams { | **Phosphorus** | 291 |
| | **Iron** | .2.7 |
| | **Sodium** | **430 |
| | **Potassium** | N.A. |
| | **Thiamin** | .08 |
| | **Riboflavin** | .23 |
| | **Niacin** | 10.3 |
| | **Ascorbic Acid** | 2 |
| unit | **Vitamin A** | 590 |

*Prepared with tuna, celery, salad dressing (mayonnaise type), pickle, onion, and egg.
**Value for product without salt added.

# BEEF AND VEAL

BEEF AND VEGETABLE STEW
BEEF HEART
BEEF LIVER
BEEF POT PIE
BEEF POT ROAST
CANNED CORN BEEF AND
CORNED BEEF HASH
CHILI CON CARNE
CHOP SUEY WITH BEEF
DRIED CHIPPED BEEF
GROUND BEEF
ROAST BEEF
ROUND STEAK
SIRLOIN STEAK
VEAL CUTLET
VEAL RIB

SEE ALSO SAUSAGES

NOTE: On all cuts of meat which are sold with a layer of fat on them, the fat has been removed to within approximately ½ inch of the lean. Deposits of fat within the cut have not been removed.

# BEEF & VEGETABLE
## STEW

| | grams | Weight: | 245 | 245 |
|---|---|---|---|---|

| | | home recipe, lean beef chunks | canned |
|---|---|---|---|
| % | Water | 82 | 82.5 |
| grams | Calories | 220 | 194 |
| | Protein | 16 | 14.2 |
| | Fat | 11 | 7.6 |
| | Carbohydrate | 15 | 17.4 |
| | Calcium | 29 | 29 |
| milligrams | Phosphorus | 184 | 110 |
| | Iron | 2.9 | 2.2 |
| | Sodium | *91 | 1,007 |
| | Potassium | 613 | 426 |
| | Thiamin | .15 | .07 |
| | Riboflavin | .17 | .12 |
| | Niacin | 4.7 | 2.5 |
| | Ascorbic Acid | 17 | 7 |
| unit | Vitamin A | 2,400 | 2,380 |

*Applies to product prepared without added salt. With salt, approximate value for 100 g. of stew is 119 mg; for 1 cup 292 mg.

# BEEF HEART

grams **Weight:** 85

lean braised

| | | |
|---|---|---|
| % | **Water** | 61 |
| | **Calories** | 160 |
| grams | **Protein** | 27 |
| | **Fat** | 5 |
| | **Carbohydrate** | 1 |
| | **Calcium** | 5 |
| milligrams | **Phosphorus** | 154 |
| | **Iron** | 5.0 |
| | **Sodium** | 87 |
| | **Potassium** | 197 |
| | **Thiamin** | .21 |
| | **Riboflavin** | 1.04 |
| | **Niacin** | 6.5 |
| | **Ascorbic Acid** | 1 |
| unit | **Vitamin A** | 20 |

103

# BEEF LIVER

| | Weight: | 85 |
|---|---|---|
| grams | | |

fried*
slice 6½ by
2⅜ by ⅜ in.

| | | |
|---|---|---|
| % | **Water** | 56 |
| | **Calories** | 195 |
| grams { | **Protein** | 22 |
| | **Fat** | 9 |
| | **Carbohydrate** | 5 |
| | **Calcium** | 9 |
| milligrams { | **Phosphorus** | 405 |
| | **Iron** | 75 |
| | **Sodium** | 156 |
| | **Potassium** | 323 |
| | **Thiamin** | .22 |
| | **Riboflavin** | 3.56 |
| | **Niacin** | 14.0 |
| | **Ascorbic Acid** | 23 |
| unit | **Vitamin A** | **45,390 |

*Regular type margarine used.
**Value varies widely.

# BEEF POT PIE

| | grams | **Weight:** | 210 |

Baked*
Home recipe

| | | |
|---|---|---|
| % | **Water** | 55 |
| | **Calories** | 515 |
| grams { | **Protein** | 21 |
| | **Fat** | 30 |
| | **Carbohydrate** | 39 |
| | **Calcium** | 29 |
| milligrams { | **Phosphorus** | 149 |
| | **Iron** | 3.8 |
| | **Sodium** | 596 |
| | **Potassium** | 334 |
| | **Thiamin** | .30 |
| | **Riboflavin** | .30 |
| | **Niacin** | 5.5 |
| | **Ascorbic Acid** | 6 |
| unit | **Vitamin A** | 1,720 |

*Crust made with vegetable shortening and enriched flour.

105

# BEEF POT ROAST

| | grams | grams |
|---|---|---|
| **Weight:** | 85 | 72 |
| | OZ. 3 | OZ. 2.5 |
| | lean and fat | lean |

| % | | | |
|---|---|---|---|
| | **Water** | 53 | 62 |
| | **Calories** | 245 | 140 |
| grams | **Protein** | 23 | 22 |
| | **Fat** | 16 | 5 |
| | **Carbohydrate** | 0 | 0 |
| | **Calcium** | 10 | 10 |
| milligrams | **Phosphorus** | 114 | 108 |
| | **Iron** | 2.9 | 2.7 |
| | **Sodium** | 40 | 45 |
| | **Potassium** | 184 | 176 |
| | **Thiamin** | .04 | .04 |
| | **Riboflavin** | .18 | .17 |
| | **Niacin** | 3.6 | 3.3 |
| | **Ascorbic Acid** | N.A. | N.A. |
| unit | **Vitamin A** | 30 | 10 |

106

# CANNED
# CORNED BEEF
## CORNED BEEF HASH

| | grams | Weight: | 85 | 220 |
|---|---|---|---|---|

| | | corned beef, canned | corned beef hash canned |
|---|---|---|---|
| % | Water | 59 | 67 |
| | Calories | 185 | 400 |
| grams { | Protein | 22 | 19 |
| | Fat | 10 | 25 |
| | Carbohydrate | 0 | 24 |
| | Calcium | 17 | 29 |
| milligrams { | Phosphorus | 90 | 147 |
| | Iron | 3.7 | 4.4 |
| | Sodium | N.A. | 1,188 |
| | Potassium | N.A. | 440 |
| | Thiamin | .01 | .02 |
| | Riboflavin | .20 | .20 |
| | Niacin | 2.9 | 4.6 |
| | Ascorbic Acid | N.A. | N.A. |
| unit | Vitamin A | N.A. | N.A. |

# CHILI CON CARNE
## WITH BEANS

grams **Weight:** 255

canned

| | % | |
|---|---|---|
| **Water** | 72 | |

| | grams | |
|---|---|---|
| **Calories** | 340 | |
| **Protein** | 19 | |
| **Fat** | 16 | |
| **Carbohydrate** | 31 | |
| **Calcium** | 82 | |

| | milligrams | |
|---|---|---|
| **Phosphorus** | 321 | |
| **Iron** | 4.3 | |
| **Sodium** | 1,354 | |
| **Potassium** | 594 | |
| **Thiamin** | .08 | |
| **Riboflavin** | .18 | |
| **Niacin** | 33. | |
| **Ascorbic Acid** | N.A. | |

| | unit | |
|---|---|---|
| **Vitamin A** | 150 | |

108

# CHOP SUEY
## WITH BEEF

grams **Weight:** 250

Homemade

| | |
|---|---|
| % Water | 75 |
| Calories | 300 |
| Protein | 26 |
| Fat | 17 |
| Carbohydrate | 13 |
| Calcium | 60 |
| Phosphorus | 248 |
| Iron | 4.8 |
| Sodium | 1,053 |
| Potassium | 425 |
| Thiamin | .28 |
| Riboflavin | .38 |
| Niacin | 5.0 |
| Ascorbic Acid | 33 |
| Vitamin A | 600 |

grams { Protein, Fat, Carbohydrate, Calcium

milligrams { Phosphorus ... Ascorbic Acid

unit Vitamin A

# DRIED
# CHIPPED BEEF

grams **Weight:** 71

| | | |
|---|---|---|
| % | Water | 48 |
| | Calories | 145 |
| grams | Protein | 24 |
| | Fat | 4 |
| | Carbohydrate | 0 |
| | Calcium | 14 |
| milligrams | Phosphorus | 287 |
| | Iron | 3.6 |
| | Sodium | 3,053 |
| | Potassium | 142 |
| | Thiamin | .05 |
| | Riboflavin | .23 |
| | Niacin | 2.7 |
| | Ascorbic Acid | 0 |
| unit | Vitamin A | N.A. |

# GROUND BEEF

| grams | **Weight:** | 85 | 82 |
|---|---|---|---|

| | | lean with 10% fat* | lean with 21% fat** |
|---|---|---|---|
| % | **Water** | 60 | 54 |
| | **Calories** | 185 | 235 |
| | **Protein** | 23 | 20 |
| | **Fat** | 10 | 17 |
| | **Carbohydrate** | 0 | 0 |
| | **Calcium** | 10 | 9 |
| | **Phosphorus** | 196 | 159 |
| | **Iron** | 3.0 | 2.6 |
| | **Sodium** | 57 | 48.2 |
| | **Potassium** | 261 | 221 |
| | **Thiamin** | .08 | .07 |
| | **Riboflavin** | .20 | .17 |
| | **Niacin** | 5.1 | 4.4 |
| | **Ascorbic Acid** | N.A. | N.A. |
| unit | **Vitamin A** | 20 | 30 |

*3 oz. or 3 x 5/8 patty.
**2.9 oz. or 3 x 5/8 patty.

# ROAST BEEF
## (FAT)

| | grams | Weight: | 85 | 51 |
|---|---|---|---|---|

| | | Relatively<br>fat<br>lean & fat | Relatively<br>fat<br>lean only |
|---|---|---|---|
| % | Water | 40 | 57 |
| | Calories | 375 | 125 |
| grams | Protein | 17 | 14 |
| | Fat | 33 | 7 |
| | Carbohydrate | 0 | 0 |
| | Calcium | 8 | 6 |
| milligrams | Phosphorus | 158 | 131 |
| | Iron | 2.2 | 1.8 |
| | Sodium | 49 | N.A. |
| | Potassium | 189 | 161 |
| | Thiamin | .05 | .04 |
| | Riboflavin | .13 | .11 |
| | Niacin | 3.1 | 2.6 |
| | Ascorbic Acid | N.A. | N.A. |
| unit | Vitamin A | 70 | 10 |

# ROAST BEEF
## (LEAN)

| | grams | Weight: | 85 | 78 |
|---|---|---|---|---|

| | | Relatively lean lean & fat | Relatively lean lean only |
|---|---|---|---|
| % | Water | 62 | 65 |
| grams | Calories | 165 | 125 |
| | Protein | 25 | 24 |
| | Fat | 7 | 3 |
| | Carbohydrate | 0 | 0 |
| | Calcium | 11 | 10 |
| milligrams | Phosphorus | 208 | 199 |
| | Iron | 3.2 | 3.0 |
| | Sodium | 61 | N.A. |
| | Potassium | 279 | 268 |
| | Thiamin | .06 | .06 |
| | Riboflavin | .19 | .18 |
| | Niacin | 4.5 | 4.3 |
| | Ascorbic Acid | N.A. | N.A. |
| unit | Vitamin A | 10 | Trace |

# ROUND STEAK

| | grams | Weight: | 85 | 68 |
|---|---|---|---|---|

|  | | relatively lean; braised: lean & fat | relatively lean; braised: lean only |
|---|---|---|---|
| % | Water | 55 | 61 |
| | Calories | 220 | 130 |
| grams | Protein | 24 | 21 |
| | Fat | 13 | 4 |
| | Carbohydrate | 0 | 0 |
| | Calcium | 10 | 9 |
| milligrams | Phosphorus | 213 | 182 |
| | Iron | 3.0 | 2.5 |
| | Sodium | 60 | 65 |
| | Potassium | 272 | 238 |
| | Thiamin | .07 | .05 |
| | Riboflavin | .19 | .16 |
| | Niacin | 4.8 | 4.1 |
| | Ascorbic Acid | N.A. | N.A. |
| unit | Vitamin A | 20 | 10 |

# SIRLOIN STEAK

| grams | Weight: | 85 | 56 |
|---|---|---|---|
| | | relatively fat broiled: lean & fat | relatively fat broiled: lean only |

| | | | |
|---|---|---|---|
| % | Water | 44 | 59 |
| | Calories | 330 | 115 |
| grams | Protein | 20 | 18 |
| | Fat | 27 | 4 |
| | Carbohydrate | 0 | 0 |
| | Calcium | 9 | 7 |
| milligrams | Phosphorus | 162 | 146 |
| | Iron | 2.5 | 2.2 |
| | Sodium | 48 | 67 |
| | Potassium | 220 | 202 |
| | Thiamin | .05 | .05 |
| | Riboflavin | .15 | .14 |
| | Niacin | 4.0 | 3.6 |
| | Ascorbic Acid | N.A. | N.A. |
| unit | Vitamin A | 50 | 10 |

# VEAL CUTLET

grams   **Weight:**               85

braised or
broiled*

| % | **Water** | 60 |
|---|---|---|
| | **Calories** | 185 |
| | **Protein** | 23 |
| | **Fat** | 9 |
| | **Carbohydrate** | 0 |
| | **Calcium** | 9 |
| | **Phosphorus** | 196 |
| | **Iron** | 2.7 |
| | **Sodium** | 41 |
| | **Potassium** | 258 |
| | **Thiamin** | .06 |
| | **Riboflavin** | .21 |
| | **Niacin** | 4.6 |
| | **Ascorbic Acid** | N.A. |
| unit | **Vitamin A** | N.A. |

*Bone removed.

116

# VEAL RIB

grams **Weight:** 85

Roasted

| | | |
|---|---|---|
| % | **Water** | 55 |
| | **Calories** | 230 |
| grams | **Protein** | 23 |
| | **Fat** | 14 |
| | **Carbohydrate** | 0 |
| | **Calcium** | 10 |
| | **Phosphorus** | 211 |
| | **Iron** | 2.9 |
| | **Sodium** | 57 |
| milligrams | **Potassium** | 259 |
| | **Thiamin** | .11 |
| | **Riboflavin** | .26 |
| | **Niacin** | 6.6 |
| | **Ascorbic Acid** | N.A. |
| unit | **Vitamin A** | N.A. |

# LAMB

LAMB CHOPS
LAMB LEG ROAST
LAMB SHOULDER ROAST

NOTE: On all cuts of meat which are sold with a layer of fat on them, the fat has been removed to within approximately ½ inch of the lean. Deposits of fat within the cut have not been removed.

# LAMB CHOPS
## (BROILED)

| grams | **Weight:** | 95 | 65 |
|---|---|---|---|

Loin chops lean & fat*    Loin chops lean only**

| | | | |
|---|---|---|---|
| % | **Water** | 47.0 | 62.1 |
| | **Calories** | 341 | 122 |
| grams | **Protein** | 20.9 | 18.3 |
| | **Fat** | 27.9 | 4.9 |
| | **Carbohydrate** | 0 | 0 |
| | **Calcium** | 9 | 8 |
| | **Phosphorus** | 163 | 142 |
| | **Iron** | 1.2 | 1.3 |
| milligrams | **Sodium** | 51 | 45 |
| | **Potassium** | 234 | 205 |
| | **Thiamin** | .11 | .10 |
| | **Riboflavin** | .22 | .18 |
| | **Niacin** | 4.8 | 4.0 |
| | **Ascorbic Acid** | N.A. | N.A. |
| unit | **Vitamin A** | N.A. | N.A. |

\* 3.4 oz. or cut 3 per lb.
\*\*2.3 oz. or cut 3 per lb.

119

# LEG ROAST
# LAMB
## (ROASTED)

| grams | Weight: | 85 | 71 |
|---|---|---|---|

| | lean & fat | lean only |
|---|---|---|

| % | | | |
|---|---|---|---|
| | **Water** | 54 | 62 |
| | **Calories** | 235 | 130 |
| | **Protein** | 22 | 20 |
| grams | **Fat** | 16 | 5 |
| | **Carbohydrate** | 0 | 0 |
| | **Calcium** | 9 | 9 |
| | **Phosphorus** | 177 | 169 |
| | **Iron** | 1.4 | 1.4 |
| | **Sodium** | 53 | 60 |
| milligrams | **Potassium** | 241 | 227 |
| | **Thiamin** | .13 | .12 |
| | **Riboflavin** | .23 | .21 |
| | **Niacin** | 4.7 | 4.4 |
| | **Ascorbic Acid** | N.A. | N.A. |
| unit | **Vitamin A** | N.A. | N.A. |

## SHOULDER ROAST
# LAMB
## (ROASTED)

| grams | Weight: | 85 | 64 |
|---|---|---|---|

| | lean & fat | lean only |
|---|---|---|

| % | Water | 50 | 61 |
|---|---|---|---|
| | **Calories** | 285 | 130 |
| | **Protein** | 18 | 17 |
| grams | **Fat** | 23 | 6 |
| | **Carbohydrate** | 0 | 0 |
| | **Calcium** | 9 | 8 |
| | **Phosphorus** | 146 | 140 |
| | **Iron** | 1.0 | 1.0 |
| | **Sodium** | 45 | 56 |
| milligrams | **Potassium** | 206 | 193 |
| | **Thiamin** | .11 | .10 |
| | **Riboflavin** | .20 | .18 |
| | **Niacin** | 4.0 | 3.7 |
| | **Ascorbic Acid** | N.A. | N.A. |
| unit | **Vitamin A** | N.A. | N.A. |

# PORK

BACON
BACON, CANADIAN STYLE
HAM
PORK CHOP
PORK LUNCHEON MEAT
PORK ROAST
PORK SHOULDER
DEVILED HAM
SEE ALSO SAUSAGES

NOTE:   On all cuts of meat which are sold with a layer of fat on them, the fat has been removed to within approximately ½ inch of the lean. Deposits of fat within the cut have not been removed.

# BACON

grams **Weight:** 15

Broiled or
fried crisp*

| | Broiled or fried crisp* |
|---|---|
| % Water | 8 |
| Calories | 85 |
| Protein | 4 |
| Fat | 8 |
| Carbohydrate | Trace |
| Calcium | 2 |
| Phosphorus | 34 |
| Iron | .5 |
| Sodium | 153 |
| Potassium | 35 |
| Thiamin | .08 |
| Riboflavin | .05 |
| Niacin | .8 |
| Ascorbic Acid | N.A. |
| Vitamin A | 0 |

*20 slices per lb.

# CANADIAN STYLE
# BACON

grams **Weight:** 21

broiled or
fried*

| | % | |
|---|---|---|
| Water | 49.9 | |

| | | |
|---|---|---|
| Calories | 58 | |
| Protein | 5.7 | |
| Fat | 3.7 | |
| Carbohydrate | .1 | |
| Calcium | 3 | |
| Phosphorus | 46 | |
| Iron | .9 | |
| Sodium | 537 | |
| Potassium | 91 | |
| Thiamin | .19 | |
| Riboflavin | .04 | |
| Niacin | 22.7 | |
| Ascorbic Acid | N.A. | |
| Vitamin A | 0 | |

grams { Protein, Fat, Carbohydrate }
milligrams { Calcium ... Ascorbic Acid }
unit

*Uncooked slice 3⅜" diam., 3/16" thick.

124

# HAM
## CURED AND COOKED

grams **Weight:** 85

light cure,
lean & fat
roasted

| | | |
|---|---|---|
| % | **Water** | 54 |
| | **Calories** | 245 |
| grams | **Protein** | 18 |
| | **Fat** | 19 |
| | **Carbohydrate** | 0 |
| | **Calcium** | 8 |
| milligrams | **Phosphorus** | 146 |
| | **Iron** | 2.2 |
| | **Sodium** | 271 |
| | **Potassium** | 199 |
| | **Thiamin** | .40 |
| | **Riboflavin** | .15 |
| | **Niacin** | 3.1 |
| | **Ascorbic Acid** | N.A. |
| unit | **Vitamin A** | 0 |

# LOIN
# PORK CHOP
# (COOKED)

| grams | Weight: | 78 | 56 |
|---|---|---|---|

| | lean & fat | lean only |
|---|---|---|

| % | | | |
|---|---|---|---|
| % | **Water** | 42 | 53 |
| | **Calories** | 305 | 150 |
| | **Protein** | 19 | 17 |
| grams | **Fat** | 25 | 9 |
| | **Carbohydrate** | 0 | 0 |
| | **Calcium** | 9 | 7 |
| | **Phosphorus** | 209 | 181 |
| | **Iron** | 2.7 | 2.2 |
| | **Sodium** | 47 | 42 |
| milligrams | **Potassium** | 216 | 192 |
| | **Thiamin** | .75 | .63 |
| | **Riboflavin** | .22 | .18 |
| | **Niacin** | 4.5 | 3.8 |
| | **Ascorbic Acid** | N.A. | N.A. |
| unit | **Vitamin A** | 0 | 0 |

126

# PORK
## LUNCHEON MEAT

| | grams | Weight: | 28 | 60 |
|---|---|---|---|---|

| | | broiled ham slice | canned spiced or unspiced |
|---|---|---|---|
| % | **Water** | 59 | 55 |
| | **Calories** | 65 | 175 |
| | **Protein** | 5 | 9 |
| grams | **Fat** | 5 | 15 |
| | **Carbohydrate** | 0 | 1 |
| | **Calcium** | 3 | 5 |
| | **Phosphorus** | 47 | 65 |
| | **Iron** | .8 | 1.3 |
| | **Sodium** | 396 | 366 |
| milligrams | **Potassium** | N.A. | 133 |
| | **Thiamin** | .12 | .19 |
| | **Riboflavin** | .04 | .13 |
| | **Niacin** | .7 | 1.8 |
| | **Ascorbic Acid** | N.A. | N.A. |
| unit | **Vitamin A** | 0 | 0 |

# PORK ROAST
## (OVEN COOKED)

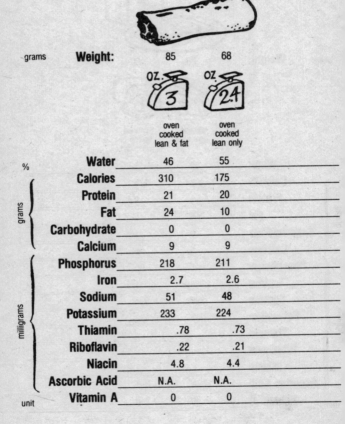

| | grams Weight: | 85 | 68 |
|---|---|---|---|
| | | OZ. 3 | OZ. 2.4 |
| | | oven cooked lean & fat | oven cooked lean only |
| % | Water | 46 | 55 |
| | Calories | 310 | 175 |
| grams | Protein | 21 | 20 |
| | Fat | 24 | 10 |
| | Carbohydrate | 0 | 0 |
| | Calcium | 9 | 9 |
| milligrams | Phosphorus | 218 | 211 |
| | Iron | 2.7 | 2.6 |
| | Sodium | 51 | 48 |
| | Potassium | 233 | 224 |
| | Thiamin | .78 | .73 |
| | Riboflavin | .22 | .21 |
| | Niacin | 4.8 | 4.4 |
| | Ascorbic Acid | N.A. | N.A. |
| unit | Vitamin A | 0 | 0 |

# PORK SHOULDER

| | grams | Weight: | 85 | 85 |
|---|---|---|---|---|

| | | simmered lean & fat | simmered lean only |
|---|---|---|---|
| % | **Water** | 46 | 53.9 |
| | **Calories** | 320 | 207 |
| grams { | **Protein** | 20 | 23.6 |
| | **Fat** | 26 | 11.7 |
| | **Carbohydrate** | 0 | 0 |
| | **Calcium** | 9 | 10 |
| | **Phosphorus** | 118 | 185 |
| | **Iron** | 2.6 | 3.1 |
| milligrams { | **Sodium** | 698 | 845 |
| | **Potassium** | 158 | 264 |
| | **Thiamin** | .46 | .54 |
| | **Riboflavin** | .21 | .21 |
| | **Niacin** | 4.1 | 4.3 |
| | **Ascorbic Acid** | N.A. | N.A. |
| unit | **Vitamin A** | 0 | 0 |

# DEVILED HAM

grams **Weight:** 13

canned

| | | |
|---|---|---|
| % | **Water** | 51 |
| | **Calories** | 45 |
| grams { | **Protein** | 2 |
| | **Fat** | 4 |
| | **Carbohydrate** | 0 |
| | **Calcium** | 1 |
| | **Phosphorus** | 12 |
| | **Iron** | .3 |
| | **Sodium** | 128 |
| milligrams { | **Potassium** | N.A. |
| | **Thiamin** | .02 |
| | **Riboflavin** | .01 |
| | **Niacin** | .2 |
| | **Ascorbic Acid** | N.A. |
| unit | **Vitamin A** | 0 |

# POULTRY

CHICKEN A LA KING
CHICKEN AND NOODLES
CHICKEN BACK
CHICKEN BREAST
CHICKEN, BROILER
CHICKEN, CANNED, BONELESS
CHICKEN CHOW MEIN
CHICKEN DRUMSTICK
CHICKEN NECK
CHICKEN POT PIE
CHICKEN THIGH
CHICKEN WING
TURKEY, ROASTED
TURKEY, CHOPPED, DICED

SEE ALSO SAUSAGES (POTTED MEAT)

# CHICKEN A LA KING

| | | |
|---|---|---|
| grams | **Weight:** | 245 |

cooked
(home recipe)

| | | |
|---|---|---|
| % | **Water** | 68 |
| | **Calories** | 470 |
| grams | **Protein** | 27 |
| | **Fat** | 34 |
| | **Carbohydrate** | 12 |
| | **Calcium** | 127 |
| milligrams | **Phosphorus** | 358 |
| | **Iron** | 25 |
| | **Sodium** | 760 |
| | **Potassium** | 404 |
| | **Thiamin** | .10 |
| | **Riboflavin** | .42 |
| | **Niacin** | 5.4 |
| | **Ascorbic Acid** | 12 |
| unit | **Vitamin A** | 1,130 |

# CHICKEN AND NOODLES

| grams | **Weight:** | 240 |
|---|---|---|

chicken &
noodles, cooked
(home recipe)

| | | |
|---|---|---|
| % | **Water** | 71 |
| | **Calories** | 365 |
| grams | **Protein** | 22 |
| | **Fat** | 18 |
| | **Carbohydrate** | 26 |
| | **Calcium** | 26 |
| | **Phosphorus** | 247 |
| | **Iron** | 2.2 |
| | **Sodium** | 600 |
| milligrams | **Potassium** | 149 |
| | **Thiamin** | .05 |
| | **Riboflavin** | .17 |
| | **Niacin** | 4.3 |
| | **Ascorbic Acid** | Trace |
| unit | **Vitamin A** | 430 |

133

# CHICKEN
## BACK

| | grams | **Weight:** | 60 |

fried in
vegetable short-
ening - bones
not removed*

| | | |
|---|---|---|
| % | **Water** | 40.5 |
| grams { | **Calories** | 139 |
| | **Protein** | 12.1 |
| | **Fat** | 8.5 |
| | **Carbohydrate** | 2.7 |
| | **Calcium** | 6 |
| milligrams { | **Phosphorus** | 105 |
| | **Iron** | 1.1 |
| | **Sodium** | N.A. |
| | **Potassium** | N.A. |
| | **Thiamin** | .03 |
| | **Riboflavin** | .20 |
| | **Niacin** | 2.7 |
| | **Ascorbic Acid** | N.A. |
| unit | **Vitamin A** | 160 |

*Refuse: bones 33%

134

# CHICKEN BREAST

grams  **Weight:**  79

Breast, fried
bones
removed*

| | | |
|---|---|---|
| % | **Water** | 58 |
| | **Calories** | 160 |
| grams | **Protein** | 26 |
| | **Fat** | 5 |
| | **Carbohydrate** | 1 |
| | **Calcium** | 9 |
| milligrams | **Phosphorus** | 218 |
| | **Iron** | 1.3 |
| | **Sodium** | N.A. |
| | **Potassium** | N.A. |
| | **Thiamin** | .04 |
| | **Riboflavin** | .17 |
| | **Niacin** | 11.6 |
| | **Ascorbic Acid** | N.A. |
| unit | **Vitamin A** | 70 |

*Fried in vegetable shortening.

# CHICKEN
## (½) BROILER

grams | **Weight:** | 176

Broiled,
bones
removed

| | | |
|---|---|---|
| % | **Water** | 71 |
| | **Calories** | 240 |
| grams | **Protein** | 42 |
| | **Fat** | 7 |
| | **Carbohydrate** | 0 |
| | **Calcium** | 16 |
| | **Phosphorus** | 355 |
| | **Iron** | 3.0 |
| | **Sodium** | 116 |
| milligrams | **Potassium** | 483 |
| | **Thiamin** | .09 |
| | **Riboflavin** | .34 |
| | **Niacin** | 15.5 |
| | **Ascorbic Acid** | N.A. |
| unit | **Vitamin A** | 160 |

# CANNED
# CHICKEN
## BONELESS

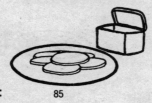

grams · **Weight:** · 85

canned,
boneless

| | | |
|---|---|---|
| % | **Water** | 65 |
| | **Calories** | 170 |
| grams | **Protein** | 18 |
| | **Fat** | 10 |
| | **Carbohydrate** | 0 |
| | **Calcium** | 18 |
| milligrams | **Phosphorus** | 210 |
| | **Iron** | 1.3 |
| | **Sodium** | N.A. |
| | **Potassium** | 117 |
| | **Thiamin** | .03 |
| | **Riboflavin** | .11 |
| | **Niacin** | 3.7 |
| | **Ascorbic Acid** | 3 |
| unit | **Vitamin A** | 200 |

# CHICKEN CHOW MEIN

| | canned, cooked | home recipe, cooked |
|---|---|---|
| **Weight:** grams | 250 | 250 |

| | canned, cooked | home recipe, cooked |
|---|---|---|
| **Water** % | 89 | 78 |
| **Calories** | 95 | 255 |
| **Protein** grams | 7 | 31 |
| **Fat** | Trace | 10 |
| **Carbohydrate** | 18 | 10 |
| **Calcium** | 45 | 58 |
| **Phosphorus** | 85 | 293 |
| **Iron** | 1.3 | 2.5 |
| **Sodium** milligrams | 725 | 718 |
| **Potassium** | 418 | 473 |
| **Thiamin** | .05 | .08 |
| **Riboflavin** | .10 | .23 |
| **Niacin** | 1.0 | 4.3 |
| **Ascorbic Acid** | 13 | 10 |
| **Vitamin A** unit | 150 | 280 |

# CHICKEN DRUMSTICK

| | | |
|---|---|---|
| grams | **Weight:** | 56 |

bones
removed*

| | | |
|---|---|---|
| % | **Water** | 55.0 |
| | **Calories** | 88 |
| | **Protein** | 12.2 |
| grams | **Fat** | 3.8 |
| | **Carbohydrate** | .4 |
| | **Calcium** | 6 |
| | **Phosphorus** | 89 |
| | **Iron** | .9 |
| | **Sodium** | N.A. |
| milligrams | **Potassium** | N.A. |
| | **Thiamin** | .03 |
| | **Riboflavin** | .15 |
| | **Niacin** | 2.7 |
| | **Ascorbic Acid** | N.A. |
| unit | **Vitamin A** | 50 |

*Fried in vegetable shortening.

# CHICKEN NECK

| | grams | **Weight:** | 60 |
|---|---|---|---|

fried in vege-
table shorten-
ing, bones not
removed*

| | | |
|---|---|---|
| % | **Water** | 50.2 |
| | **Calories** | 127 |
| grams | **Protein** | 11.7 |
| | **Fat** | 7.6 |
| | **Carbohydrate** | 2.0 |
| | **Calcium** | 5 |
| milligrams | **Phosphorus** | 102 |
| | **Iron** | 1.2 |
| | **Sodium** | N.A. |
| | **Potassium** | N.A. |
| | **Thiamin** | .04 |
| | **Riboflavin** | .18 |
| | **Niacin** | 2.5 |
| | **Ascorbic Acid** | N.A. |
| unit | **Vitamin A** | 150 |

*Refuse: bones 27%

# CHICKEN POT PIE

grams    **Weight:**    232

Baked*

| | | |
|---|---|---|
| % | **Water** | 57 |
| grams { | **Calories** | 545 |
| | **Protein** | 23 |
| | **Fat** | 31 |
| | **Carbohydrate** | 42 |
| | **Calcium** | 70 |
| milligrams { | **Phosphorus** | 232 |
| | **Iron** | 3.0 |
| | **Sodium** | 594 |
| | **Potassium** | 343 |
| | **Thiamin** | .34 |
| | **Riboflavin** | .31 |
| | **Niacin** | 5.5 |
| | **Ascorbic Acid** | 5 |
| unit | **Vitamin A** | 3;090 |

*Crust made with vegetable shortening and enriched flour.
**Piece is ⅓ of 9″ diam. pie.

# CHICKEN THIGH

grams **Weight:** 65

fried in vege-
table shorten-
ing, bones not
removed*

| | | |
|---|---|---|
| % | **Water** | 55.8 |
| | **Calories** | 122 |
| grams { | **Protein** | 15 |
| | **Fat** | 5.9 |
| | **Carbohydrate** | 1.3 |
| | **Calcium** | 7 |
| milligrams { | **Phosphorus** | 121 |
| | **Iron** | 1.2 |
| | **Sodium** | N.A. |
| | **Potassium** | N.A. |
| | **Thiamin** | .03 |
| | **Riboflavin** | .25 |
| | **Niacin** | 3.5 |
| | **Ascorbic Acid** | N.A. |
| unit | **Vitamin A** | 100 |

*Refuse: bones 21%

# CHICKEN WING

grams    **Weight:**     50

fried in vege-
table shorten-
ing, bones not
removed*

| | |
|---|---|
| %   Water | 52.6 |
| Calories | 82 |
| Protein | 8.8 |
| Fat | 4.5 |
| Carbohydrate | .8 |
| Calcium | 3 |
| Phosphorus | 72 |
| Iron | .6 |
| Sodium | N.A. |
| Potassium | N.A. |
| Thiamin | .02 |
| Riboflavin | .08 |
| Niacin | 2.1 |
| Ascorbic Acid | N.A. |
| Vitamin A | 80 |

grams — Water through Calcium
milligrams — Phosphorus through Vitamin A
unit — Vitamin A

*Refuse: bones 39%

143

# TURKEY*
## (ROASTED)

| | grams | **Weight:** | 85 | 85 |
|---|---|---|---|---|

| | Dark meat | Light meat |
|---|---|---|

| % | | | | |
|---|---|---|---|---|
| | **Water** | 61 | 62 |
| | **Calories** | 175 | 150 |
| grams | **Protein** | 26 | 28 |
| | **Fat** | 7 | 3 |
| | **Carbohydrate** | 0 | 0 |
| | **Calcium** | N.A. | N.A. |
| | **Phosphorus** | N.A. | N.A. |
| | **Iron** | 2.0 | 1.0 |
| milligrams | **Sodium** | 84 | 70 |
| | **Potassium** | 338 | 349 |
| | **Thiamin** | .03 | .04 |
| | **Riboflavin** | .20 | .12 |
| | **Niacin** | 3.6 | 9.4 |
| | **Ascorbic Acid** | N.A. | N.A. |
| unit | **Vitamin A** | N.A. | N.A. |

*Roasted, skin removed.

144

# TURKEY*
## (CHOPPED, DICED)

| | grams | **Weight:** | 140 | 85 |

** 

| | | Light & dark meat: (chopped or diced) | Light & dark meat: pieces |
|---|---|---|---|
| % | **Water** | 61 | 61 |
| | **Calories** | 265 | 160 |
| grams | **Protein** | 44 | 27 |
| | **Fat** | 9 | 5 |
| | **Carbohydrate** | 0 | 0 |
| | **Calcium** | 11 | 7 |
| milligrams | **Phosphorus** | 351 | 213 |
| | **Iron** | 2.5 | 1.5 |
| | **Sodium** | N.A. | N.A. |
| | **Potassium** | 514 | 312 |
| | **Thiamin** | .07 | .04 |
| | **Riboflavin** | .25 | .15 |
| | **Niacin** | 10.8 | 6.5 |
| | **Ascorbic Acid** | N.A. | N.A. |
| unit | **Vitamin A** | N.A. | N.A. |

*Roasted, skin removed.
**1 slice white meat, 4 by 2 by ¼″ with 2 slices dark meat, 2½ by 1⅝ by ¼″.

# SAUSAGES

BOLOGNA
BRAUNSCHWEIGER
BROWN AND SERVE
FRANKFURTER
POTTED MEAT
PORK LINK
SALAMI, DRY
SALAMI, COOKED
VIENNA SAUSAGE

# BOLOGNA

grams **Weight:** 28

| | | |
|---|---|---|
| % | **Water** | 56 |
| | **Calories** | 85 |
| grams | **Protein** | 3 |
| | **Fat** | 8 |
| | **Carbohydrate** | Trace |
| | **Calcium** | 2 |
| | **Phosphorus** | 36 |
| | **Iron** | .5 |
| milligrams | **Sodium** | N.A. |
| | **Potassium** | 65 |
| | **Thiamin** | .05 |
| | **Riboflavin** | .06 |
| | **Niacin** | .7 |
| | **Ascorbic Acid** | N.A. |
| unit | **Vitamin A** | N.A. |

# BRAUNSCHWEIGER
# SAUSAGE

| grams | Weight: | 28 |
|---|---|---|

| | | |
|---|---|---|
| % | Water | 53 |
| | Calories | 90 |
| grams | Protein | 4 |
| | Fat | 8 |
| | Carbohydrate | 1 |
| | Calcium | 3 |
| | Phosphorus | 64 |
| | Iron | 1.7 |
| | Sodium | 328 |
| milligrams | Potassium | N.A. |
| | Thiamin | .05 |
| | Riboflavin | 41 |
| | Niacin | 2.3 |
| | Ascorbic Acid | N.A. |
| unit | Vitamin A | 1,850 |

148

# BROWN AND SERVE
# SAUSAGE

grams    **Weight:**     17

browned

| | % | |
|---|---|---|
| **Water** | 40 | |

| | grams | |
|---|---|---|
| **Calories** | 70 | |
| **Protein** | 3 | |
| **Fat** | 6 | |
| **Carbohydrate** | Trace | |
| **Calcium** | N.A. | |

| | milligrams | |
|---|---|---|
| **Phosphorus** | N.A. | |
| **Iron** | N.A. | |
| **Sodium** | N.A. | |
| **Potassium** | N.A. | |
| **Thiamin** | N.A. | |
| **Riboflavin** | N.A. | |
| **Niacin** | N.A. | |
| **Ascorbic Acid** | N.A. | |

| | unit | |
|---|---|---|
| **Vitamin A** | N.A. | |

# FRANKFURTER

| grams | **Weight:** | 56 |
|---|---|---|

cooked
(reheated)

| % | **Water** | 57 |
|---|---|---|
| | **Calories** | 170 |
| grams | **Protein** | 7 |
| | **Fat** | 15 |
| | **Carbohydrate** | 1 |
| | **Calcium** | 3 |
| milligrams | **Phosphorus** | 57 |
| | **Iron** | .8 |
| | **Sodium** | N.A. |
| | **Potassium** | N.A. |
| | **Thiamin** | .08 |
| | **Riboflavin** | .11 |
| | **Niacin** | 1.4 |
| | **Ascorbic Acid** | N.A. |
| unit | **Vitamin A** | N.A. |

# POTTED MEAT
## (BEEF, CHICKEN, TURKEY)

grams | **Weight:** | 13

canned

| | | |
|---|---|---|
| % | **Water** | 61 |
| | **Calories** | 30 |
| grams | **Protein** | 2 |
| | **Fat** | 2 |
| | **Carbohydrate** | 0 |
| | **Calcium** | N.A. |
| milligrams | **Phosphorus** | N.A. |
| | **Iron** | N.A. |
| | **Sodium** | N.A. |
| | **Potassium** | N.A. |
| | **Thiamin** | Trace |
| | **Riboflavin** | .03 |
| | **Niacin** | .2 |
| | **Ascorbic Acid** | N.A. |
| unit | **Vitamin A** | N.A. |

151

# PORK LINK
# SAUSAGE

grams **Weight:** 13

cooked

| | | |
|---|---|---|
| % | **Water** | 35 |
| | **Calories** | 60 |
| | **Protein** | 2 |
| | **Fat** | 6 |
| | **Carbohydrate** | Trace |
| | **Calcium** | 1 |
| | **Phosphorus** | 21 |
| | **Iron** | .3 |
| | **Sodium** | 125 |
| | **Potassium** | 35 |
| | **Thiamin** | .10 |
| | **Riboflavin** | .04 |
| | **Niacin** | .5 |
| | **Ascorbic Acid** | N.A. |
| unit | **Vitamin A** | 0 |

grams {
milligrams {

# SALAMI
## DRY TYPE

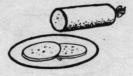

grams  **Weight:**  10

| | | |
|---|---|---|
| % | Water | 30 |
| | Calories | 45 |
| grams | Protein | 2 |
| | Fat | 4 |
| | Carbohydrate | Trace |
| | Calcium | 1 |
| | Phosphorus | 28 |
| | Iron | .4 |
| | Sodium | 180 |
| milligrams | Potassium | N.A. |
| | Thiamin | .04 |
| | Riboflavin | .03 |
| | Niacin | .05 |
| | Ascorbic Acid | N.A. |
| unit | Vitamin A | N.A. |

153

# SALAMI
## COOKED TYPE

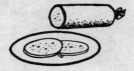

grams    **Weight:**      28

| | | |
|---|---|---|
| % | **Water** | 51 |
| | **Calories** | 90 |
| grams | **Protein** | 5 |
| | **Fat** | 7 |
| | **Carbohydrate** | Trace |
| | **Calcium** | 3 |
| | **Phosphorus** | 57 |
| | **Iron** | .7 |
| | **Sodium** | 285 |
| milligrams | **Potassium** | N.A. |
| | **Thiamin** | .07 |
| | **Riboflavin** | .07 |
| | **Niacin** | 1.2 |
| | **Ascorbic Acid** | N.A. |
| unit | **Vitamin A** | N.A. |

# VIENNA SAUSAGE

| | | |
|---|---|---|
| grams | **Weight:** | 16 |

| % | **Water** | 63 |
|---|---|---|
| | **Calories** | 40 |
| grams | **Protein** | 2 |
| | **Fat** | 3 |
| | **Carbohydrate** | Trace |
| | **Calcium** | 1 |
| milligrams | **Phosphorus** | 24 |
| | **Iron** | .3 |
| | **Sodium** | N.A. |
| | **Potassium** | N.A. |
| | **Thiamin** | .01 |
| | **Riboflavin** | .02 |
| | **Niacin** | .4 |
| | **Ascorbic Acid** | N.A. |
| unit | **Vitamin A** | N.A. |

# FRUITS

APPLES, RAW
APPLE BUTTER, CANNED
APPLESAUCE, CANNED
APRICOTS, RAW
APRICOTS, CANNED
APRICOTS, DRIED
AVOCADOS, RAW
BANANAS, RAW AND DRIED
BLACKBERRIES, RAW
BLUEBERRIES, RAW
BOYSENBERRIES, FROZEN
CANTELOUP MELON, RAW
CHERRIES, SOUR,
RAW AND CANNED
CHERRIES, SWEET,
RAW AND CANNED
CRANBERRY SAUCE,
CANNED
DATES, DRIED
FRUIT COCKTAIL, CANNED
GRAPEFRUIT, RAW
GRAPEFRUIT, CANNED
GRAPES, EUROPEAN
TYPE, RAW
HONEYDEW MELON

LEMONS, RAW
ORANGES, RAW
PAPAYAS, RAW
PEACHES, RAW
PEACHES, CANNED
PEACHES, DRIED
PEACHES, FROZEN
PEARS, RAW
PEARS, CANNED
PINEAPPLES, RAW
PINEAPPLES, CANNED
PLUMS, RAW
PLUMS, CANNED
PRUNES, DRIED
AND COOKED
RAISINS
RASPBERRIES, RAW
RASPBERRIES, FROZEN
RHUBARB, COOKED
STRAWBERRIES, RAW
STRAWBERRIES, FROZEN
TANGERINES, RAW
WATERMELON, RAW

SEE ALSO FRUIT JUICES

# APPLES

| grams | **Weight:** | 150 | 212 |
|---|---|---|---|

|  |  | 2¾ in. diam. (ab. 3 per lb. with cores) | 3¼ in. diam. (ab. 2 per lb. with cores) |
|---|---|---|---|
| % | **Water** | 84 | 84 |
|  | **Calories** | 80 | 125 |
| grams { | **Protein** | Trace | Trace |
|  | **Fat** | 1 | 1 |
|  | **Carbohydrate** | 20 | 31 |
|  | **Calcium** | 10 | 15 |
| milligrams { | **Phosphorus** | 14 | 21 |
|  | **Iron** | .4 | .6 |
|  | **Sodium** | 1 | 2 |
|  | **Potassium** | 152 | 233 |
|  | **Thiamin** | .04 | .06 |
|  | **Riboflavin** | .03 | .04 |
|  | **Niacin** | .1 | .2 |
|  | **Ascorbic Acid** | 6 | 8 |
| unit | **Vitamin A** | 120 | 190 |

# APPLE BUTTER

grams    **Weight:**      248

canned

| | | |
|---|---|---|
| % | Water | 87.8 |
| | Calories | 117 |
| grams | Protein | .2 |
| | Fat | Trace |
| | Carbohydrate | 29.2 |
| | Calcium | 15 |
| milligrams | Phosphorus | 22 |
| | Iron | 1.5 |
| | Sodium | 2 |
| | Potassium | 250 |
| | Thiamin | .02 |
| | Riboflavin | .05 |
| | Niacin | .2 |
| | Ascorbic Acid | 2 |
| unit | Vitamin A | N.A. |

# APPLESAUCE

| grams | **Weight:** | 255 | 244 |
|---|---|---|---|

| | sweetened | unsweetened |
|---|---|---|

| % | **Water** | 76 | 89 |
|---|---|---|---|
| | **Calories** | 230 | 100 |
| grams | **Protein** | 1 | Trace |
| | **Fat** | Trace | Trace |
| | **Carbohydrate** | 61 | 26 |
| | **Calcium** | 10 | 10 |
| milligrams | **Phosphorus** | 13 | 12 |
| | **Iron** | 1.3 | 1.2 |
| | **Sodium** | 5 | 5 |
| | **Potassium** | 166 | 190 |
| | **Thiamin** | .05 | .05 |
| | **Riboflavin** | .03 | .02 |
| | **Niacin** | .1 | .1 |
| | **Ascorbic Acid** | 3* | 2* |
| unit | **Vitamin A** | 100 | 100 |

*Applies to product without added ascorbic acid. For value of product with ascorbic acid, refer to label.

# APRICOTS

Weight: 114 grams

about
12 per lb.

| | |
|---|---|
| Water (%) | 85.3 |
| Calories | 55 |
| Protein (grams) | 1.1 |
| Fat | .2 |
| Carbohydrate | 13.7 |
| Calcium | 18 |
| Phosphorus (milligrams) | 25 |
| Iron | .5 |
| Sodium | 1 |
| Potassium | 301 |
| Thiamin | .03 |
| Riboflavin | .04 |
| Niacin | .6 |
| Ascorbic Acid | 11 |
| Vitamin A (unit) | 2,890 |

161

# APRICOTS
## CANNED

grams **Weight:** 258

canned in
heavy syrup
(halves &
syrup)

| | | |
|---|---|---|
| % Water | 77 | |
| Calories | 220 | |
| Protein | 2 | |
| Fat | Trace | |
| Carbohydrate | 57 | |
| Calcium | 28 | |
| Phosphorus | 39 | |
| Iron | .8 | |
| Sodium | 3 | |
| Potassium | 604 | |
| Thiamin | .05 | |
| Riboflavin | .05 | |
| Niacin | 1.0 | |
| Ascorbic Acid | 10 | |
| Vitamin A | 4,490 | |

grams { Calories, Protein, Fat, Carbohydrate, Calcium }

milligrams { Phosphorus ... Ascorbic Acid }

unit Vitamin A

# APRICOTS
## DRIED

| | grams | **Weight:** | 130 | 250 |
|---|---|---|---|---|

| | | Dried: uncooked (28 large or 37 med. halves per cup) | Dried: cooked unsweetened, fruit & liquid |
|---|---|---|---|
| % | **Water** | 25 | 76 |
| | **Calories** | 340 | 215 |
| grams | **Protein** | 7 | 4 |
| | **Fat** | 1 | 1 |
| | **Carbohydrate** | 86 | 54 |
| | **Calcium** | 87 | 55 |
| milligrams | **Phosphorus** | 140 | 88 |
| | **Iron** | 7.2 | 4.5 |
| | **Sodium** | 34 | 20 |
| | **Potassium** | 1,273 | 795 |
| | **Thiamin** | .01 | .01 |
| | **Riboflavin** | .21 | .13 |
| | **Niacin** | 4.3 | 2.5 |
| | **Ascorbic Acid** | 16 | 8 |
| unit | **Vitamin A** | 14,170 | 7,500 |

# AVOCADOS

| | grams | | |
|---|---|---|---|
| | **Weight:** | 216 | 304 |

| | | California, mid & late winter* | Florida, late summer & fall** |
|---|---|---|---|
| % | **Water** | 74 | 78 |
| | **Calories** | 370 | 390 |
| grams { | **Protein** | 5 | 4 |
| | **Fat** | 37 | 33 |
| | **Carbohydrate** | 13 | 27 |
| | **Calcium** | 22 | 30 |
| milligrams { | **Phosphorus** | 91 | 128 |
| | **Iron** | 1.3 | 1.8 |
| | **Sodium** | 9 | 12 |
| | **Potassium** | 1,303 | 1,836 |
| | **Thiamin** | .24 | .33 |
| | **Riboflavin** | .43 | .61 |
| | **Niacin** | 3.5 | 4.9 |
| | **Ascorbic Acid** | 30 | 43 |
| unit | **Vitamin A** | 630 | 880 |

*With skin and seed, 3⅛ in diam.; wt. 10 oz.
**With skin and seed 3⅝ in diam.; wt. I lb.

# BANANAS
## PEELED AND (DRIED) FLAKES

| | Weight: | 119 | 6 |
|---|---|---|---|
| grams | | | |

| | | without peel (ab. 2.6 per lb. with peel) | Banana flakes |
|---|---|---|---|
| % | **Water** | 76 | 3 |
| | **Calories** | 100 | 20 |
| grams | **Protein** | 1 | Trace |
| | **Fat** | Trace | Trace |
| | **Carbohydrate** | 26 | 5 |
| | **Calcium** | 10 | 2 |
| milligrams | **Phosphorus** | 31 | 6 |
| | **Iron** | .8 | .2 |
| | **Sodium** | 1 | Trace |
| | **Potassium** | 440 | 92 |
| | **Thiamin** | .06 | .01 |
| | **Riboflavin** | .07 | .01 |
| | **Niacin** | .8 | .2 |
| | **Ascorbic Acid** | 12 | Trace |
| unit | **Vitamin A** | 230 | 50 |

# BLACKBERRIES

grams **Weight:** 144

raw

| | | |
|---|---|---|
| % | **Water** | 85 |
| | **Calories** | 85 |
| | **Protein** | 2 |
| | **Fat** | 1 |
| | **Carbohydrate** | 19 |
| | **Calcium** | 46 |
| | **Phosphorus** | 27 |
| | **Iron** | 1.3 |
| | **Sodium** | 1 |
| | **Potassium** | 245 |
| | **Thiamin** | .04 |
| | **Riboflavin** | .06 |
| | **Niacin** | .6 |
| | **Ascorbic Acid** | 30 |
| unit | **Vitamin A** | 290 |

grams {Protein, Fat, Carbohydrate}
milligrams {Calcium ... Vitamin A}

166

# BLUEBERRIES

grams **Weight:** 145

raw

| | | |
|---|---|---|
| % | **Water** | 83 |
| grams | **Calories** | 90 |
| | **Protein** | 1 |
| | **Fat** | 1 |
| | **Carbohydrate** | 22 |
| | **Calcium** | 22 |
| milligrams | **Phosphorus** | 19 |
| | **Iron** | 1.5 |
| | **Sodium** | 1 |
| | **Potassium** | 117 |
| | **Thiamin** | .04 |
| | **Riboflavin** | .09 |
| | **Niacin** | .7 |
| | **Ascorbic Acid** | 20 |
| unit | **Vitamin A** | 150 |

167

# BOYSENBERRIES
## FROZEN, UNSWEETENED

grams **Weight:** 126

frozen, un-
sweetened

| | | |
|---|---|---|
| % | **Water** | 86.8 |
| | **Calories** | 60 |
| | **Protein** | 1.5 |
| grams | **Fat** | .4 |
| | **Carbohydrate** | 14.4 |
| | **Calcium** | 32 |
| | **Phosphorus** | 30 |
| | **Iron** | 2 |
| | **Sodium** | 1 |
| milligrams | **Potassium** | 193 |
| | **Thiamin** | .03 |
| | **Riboflavin** | .16 |
| | **Niacin** | 1.3 |
| | **Ascorbic Acid** | 16 |
| unit | **Vitamin A** | 210 |

# CANTALOUPE MELON

| | | |
|---|---|---|
| grams | **Weight:** | 477 |

5″ diam.
2⅓ lb. with
rind

| | | |
|---|---|---|
| % | **Water** | 91 |
| | **Calories** | 80 |
| grams { | **Protein** | 2 |
| | **Fat** | Trace |
| | **Carbohydrate** | 20 |
| | **Calcium** | 38 |
| milligrams { | **Phosphorus** | 44 |
| | **Iron** | 1.1 |
| | **Sodium** | 33 |
| | **Potassium** | 682 |
| | **Thiamin** | .11 |
| | **Riboflavin** | .08 |
| | **Niacin** | 1.6 |
| | **Ascorbic Acid** | 90 |
| unit | **Vitamin A** | 9,240 |

# CHERRIES
## SOUR RED

| | grams | **Weight:** | 155 | 244 |
|---|---|---|---|---|

| | | raw, pitted | pitted, water pack |
|---|---|---|---|
| % | **Water** | 83.7 | 88 |
| | **Calories** | 90 | 105 |
| grams | **Protein** | 1.9 | 2 |
| | **Fat** | .5 | Trace |
| | **Carbohydrate** | 22.2 | 26 |
| | **Calcium** | 34 | 37 |
| milligrams | **Phosphorus** | 29 | 32 |
| | **Iron** | .6 | .7 |
| | **Sodium** | 3 | 5 |
| | **Potassium** | 296 | 317 |
| | **Thiamin** | .08 | .07 |
| | **Riboflavin** | .09 | .05 |
| | **Niacin** | .6 | .5 |
| | **Ascorbic Acid** | 16 | 12 |
| unit | **Vitamin A** | 1,550 | 1,660 |

# CHERRIES
## SWEET

| grams | Weight: | 68 | 257 |
|---|---|---|---|

| | | raw | canned, pitted |
|---|---|---|---|
| % | Water | 80 | 78 |
| | Calories | 45 | 208 |
| grams | Protein | 1 | 2.3 |
| | Fat | Trace | .5 |
| | Carbohydrate | 12 | 52.7 |
| | Calcium | 15 | 39 |
| milligrams | Phosphorus | 13 | 33 |
| | Iron | .3 | .8 |
| | Sodium | 1 | 3 |
| | Potassium | 129 | 324 |
| | Thiamin | .03 | .05 |
| | Riboflavin | .04 | .05 |
| | Niacin | .3 | .5 |
| | Ascorbic Acid | 7 | 8 |
| unit | Vitamin A | 70 | 150 |

# CRANBERRY SAUCE
## CANNED

| | grams | **Weight:** | 277 |

sweetened

| | | |
|---|---|---|
| % | **Water** | 62 |
| | **Calories** | 405 |
| grams | **Protein** | Trace |
| | **Fat** | 1 |
| | **Carbohydrate** | 104 |
| | **Calcium** | 17 |
| milligrams | **Phosphorus** | 11 |
| | **Iron** | .6 |
| | **Sodium** | 3 |
| | **Potassium** | 83 |
| | **Thiamin** | .03 |
| | **Riboflavin** | .03 |
| | **Niacin** | .1 |
| | **Ascorbic Acid** | 6 |
| unit | **Vitamin A** | 60 |

# DATES

| | grams | | |
|---|---|---|---|
| | **Weight:** | 80 | 178 |
| | | 10 whole, without pits | 1 chopped |

| % | Water | 23 | 23 |
|---|---|---|---|
| | Calories | 220 | 490 |
| | Protein | 2 | 4 |
| grams | Fat | Trace | 1 |
| | Carbohydrate | 58 | 130 |
| | Calcium | 47 | 105 |
| | Phosphorus | 50 | 112 |
| | Iron | 2.4 | 5.3 |
| | Sodium | 1 | 2 |
| milligrams | Potassium | 518 | 1,153 |
| | Thiamin | .07 | .16 |
| | Riboflavin | .08 | .18 |
| | Niacin | 1.8 | 3.9 |
| | Ascorbic Acid | 0 | 0 |
| unit | Vitamin A | 40 | 90 |

# FRUIT COCKTAIL

| grams | **Weight:** | 255 |

in heavy
syrup

| % | **Water** | 80 |
|---|---|---|
| | **Calories** | 195 |
| grams | **Protein** | 1 |
| | **Fat** | Trace |
| | **Carbohydrate** | 50 |
| | **Calcium** | 23 |
| | **Phosphorus** | 31 |
| | **Iron** | 1.0 |
| | **Sodium** | 13 |
| milligrams | **Potassium** | 411 |
| | **Thiamin** | .05 |
| | **Riboflavin** | .03 |
| | **Niacin** | 1.0 |
| | **Ascorbic Acid** | 5 |
| unit | **Vitamin A** | 360 |

174

# GRAPEFRUIT

| | grams | **Weight:** | 241 | 241 |
|---|---|---|---|---|

| | | medium*<br>pink or red | medium**<br>white |
|---|---|---|---|

| | | | |
|---|---|---|---|
| % | **Water** | 89 | 89 |
| | **Calories** | 50 | 45 |
| grams { | **Protein** | 1 | 1 |
| | **Fat** | Trace | Trace |
| | **Carbohydrate** | 13 | 12 |
| | **Calcium** | 20 | 19 |
| milligrams { | **Phosphorus** | 20 | 19 |
| | **Iron** | .5 | .5 |
| | **Sodium** | 1 | 1 |
| | **Potassium** | 166 | 159 |
| | **Thiamin** | .05 | .05 |
| | **Riboflavin** | .02 | .02 |
| | **Niacin** | .2 | .2 |
| | **Ascorbic Acid** | 44 | 44 |
| unit | **Vitamin A** | 540 | 10 |

*Weight includes peel and membranes between sections. Without these parts,
the weight of the edible portion is 123g for pink & red and 118g for white.
**3¾ in. diam. (about 1 lb. 1 oz.)

# GRAPEFRUIT

grams **Weight:** 254

sections with
syrup

| % | Water | 81 | |
|---|---|---|---|
| grams { | **Calories** | 180 | |
| | **Protein** | 2 | |
| | **Fat** | Trace | |
| | **Carbohydrate** | 45 | |
| | **Calcium** | 33 | |
| milligrams { | **Phosphorus** | 36 | |
| | **Iron** | .8 | |
| | **Sodium** | 3 | |
| | **Potassium** | 343 | |
| | **Thiamin** | .08 | |
| | **Riboflavin** | .05 | |
| | **Niacin** | .5 | |
| | **Ascorbic Acid** | 76 | |
| unit | **Vitamin A** | 30 | |

# GRAPES

| grams | Weight: | 50 | 60 |
|---|---|---|---|
| | | Thompson seedless | Tokay and Emperor seeded types* |
| % | **Water** | 81 | 81 |
| | **Calories** | 35 | 40 |
| | **Protein** | Trace | Trace |
| grams | **Fat** | Trace | Trace |
| | **Carbohydrate** | 9 | 10 |
| | **Calcium** | 6 | 7 |
| | **Phosphorus** | 10 | 11 |
| | **Iron** | .2 | .2 |
| | **Sodium** | 2 | 2 |
| milligrams | **Potassium** | 87 | 99 |
| | **Thiamin** | .03 | .03 |
| | **Riboflavin** | .02 | .02 |
| | **Niacin** | .2 | .2 |
| | **Ascorbic Acid** | 2 | 2 |
| unit | **Vitamin A** | 50 | 60 |

*Weight includes seeds. Without seeds, weight of the edible portion is 57g.

# HONEYDEW MELON

grams **Weight:** 226

6½″ diam.
5¼ lb.

| | | |
|---|---|---|
| % | **Water** | 91 |
| | **Calories** | 50 |
| grams | **Protein** | 1 |
| | **Fat** | Trace |
| | **Carbohydrate** | 11 |
| | **Calcium** | 21 |
| milligrams | **Phosphorus** | 24 |
| | **Iron** | .6 |
| | **Sodium** | 18 |
| | **Potassium** | 374 |
| | **Thiamin** | .06 |
| | **Riboflavin** | .04 |
| | **Niacin** | .9 |
| | **Ascorbic Acid** | 34 |
| unit | **Vitamin A** | 60 |

178

# LEMON

grams    **Weight:**    110

Raw: 2⅛"
diam.

| | | |
|---|---|---|
| % | **Water** | 90.1 |
| | **Calories** | 20 |
| grams | **Protein** | .8 |
| | **Fat** | .2 |
| | **Carbohydrate** | 6 |
| | **Calcium** | 19 |
| milligrams | **Phosphorus** | 12 |
| | **Iron** | .4 |
| | **Sodium** | 1 |
| | **Potassium** | 102 |
| | **Thiamin** | .03 |
| | **Riboflavin** | .01 |
| | **Niacin** | .1 |
| | **Ascorbic Acid** | 39 |
| unit | **Vitamin A** | 10 |

# ORANGES

| | grams | **Weight:** | 131 | 180 |
|---|---|---|---|---|

| | whole, with-out peel & seeds 2⅝ diam. | sections without membranes |
|---|---|---|
| **Water** % | 86 | 86 |
| **Calories** | 65 | 90 |
| **Protein** | 1 | 2 |
| **Fat** | Trace | Trace |
| **Carbohydrate** | 16 | 22 |
| **Calcium** | 54 | 74 |
| **Phosphorus** | 26 | 36 |
| **Iron** | .5 | .7 |
| **Sodium** | 1 | 2 |
| **Potassium** | 263 | 360 |
| **Thiamin** | .13 | .18 |
| **Riboflavin** | .05 | .07 |
| **Niacin** | .5 | .7 |
| **Ascorbic Acid** | 66 | 90 |
| **Vitamin A** | 260 | 360 |

grams { Protein, Fat, Carbohydrate, Calcium }
milligrams { Phosphorus, Iron, Sodium, Potassium, Thiamin, Riboflavin, Niacin, Ascorbic Acid }
unit — Vitamin A

# PAPAYAS

grams **Weight:** 140

½ in. cubes

| | | |
|---|---|---|
| % | **Water** | 89 |
| | **Calories** | 55 |
| grams { | **Protein** | 1 |
| | **Fat** | Trace |
| | **Carbohydrate** | 14 |
| | **Calcium** | 28 |
| milligrams { | **Phosphorus** | 22 |
| | **Iron** | .4 |
| | **Sodium** | 4 |
| | **Potassium** | 328 |
| | **Thiamin** | .06 |
| | **Riboflavin** | .06 |
| | **Niacin** | .4 |
| | **Ascorbic Acid** | 78 |
| unit | **Vitamin A** | 2,450 |

# PEACHES

| | grams | **Weight:** | 115 | 170 |
|---|---|---|---|---|

| | | Raw: whole, 2½″ diam., peeled, pitted* | sliced |
|---|---|---|---|

| | | | |
|---|---|---|---|
| % | **Water** | 89 | 89 |
| | **Calories** | 38 | 65 |
| grams { | **Protein** | .6 | 1 |
| | **Fat** | .1 | Trace |
| | **Carbohydrate** | 9.7 | 16 |
| | **Calcium** | 9 | 15 |
| | **Phosphorus** | 19 | 32 |
| | **Iron** | .5 | .9 |
| | **Sodium** | 1 | 2 |
| milligrams { | **Potassium** | 202 | 343 |
| | **Thiamin** | .02 | .03 |
| | **Riboflavin** | .05 | .09 |
| | **Niacin** | 1.0 | 1.7 |
| | **Ascorbic Acid** | 7 | 12 |
| unit | **Vitamin A** | **1,330 | **2,260 |

*about 4 per lb. with peels and pits
**represents yellow fleshed varieties. For white fleshed varieties, value is 50 I.U. for 1 peach, 90 I.U. for 1 cup of slices.

# PEACHES
## CANNED

| | | |
|---|---|---|
| grams | **Weight:** | 256 |

| | syrup packed, canned, yellow fleshed solids & liquid (halves or slices) | water packed, canned, yellow fleshed solids & liquid |
|---|---|---|
| % | **Water** | 79 | 91 |

| | | syrup packed | water packed |
|---|---|---|---|
| % | **Water** | 79 | 91 |
| | **Calories** | 200 | 75 |
| | **Protein** | 1 | 1 |
| | **Fat** | Trace | Trace |
| | **Carbohydrate** | 51 | 20 |
| | **Calcium** | 10 | 10 |
| | **Phosphorus** | 31 | 32 |
| | **Iron** | .8 | .7 |
| | **Sodium** | 5 | 5 |
| | **Potassium** | 333 | 334 |
| | **Thiamin** | .03 | .02 |
| | **Riboflavin** | .05 | .07 |
| | **Niacin** | 1.5 | 1.5 |
| | **Ascorbic Acid** | 8 | 7 |
| unit | **Vitamin A** | 1,100 | 1,100 |

Weight: 256 grams (syrup packed), 244 grams (water packed)

grams: Protein, Fat, Carbohydrate

milligrams: Calcium, Phosphorus, Iron, Sodium, Potassium, Thiamin, Riboflavin, Niacin, Ascorbic Acid

# PEACHES
## (DRIED)

| grams | **Weight:** | 160 | 250 |
|---|---|---|---|

| | | Dried;<br>uncooked | Dried;<br>cooked, un-<br>sweetened<br>halves & juice |
|---|---|---|---|
| % | **Water** | 25 | 77 |
| grams | **Calories** | 420 | 205 |
| | **Protein** | 5 | 3 |
| | **Fat** | 1 | 1 |
| | **Carbohydrate** | 109 | 54 |
| milligrams | **Calcium** | 77 | 38 |
| | **Phosphorus** | 187 | 93 |
| | **Iron** | 9.6 | 4.8 |
| | **Sodium** | 26 | 13 |
| | **Potassium** | 1,520 | 743 |
| | **Thiamin** | .02 | .01 |
| | **Riboflavin** | .30 | .15 |
| | **Niacin** | 8.5 | 3.8 |
| | **Ascorbic Acid** | 29 | 5 |
| unit | **Vitamin A** | 6,240 | 3,050 |

# PEACHES
## FROZEN, CANNED

| | grams | **Weight:** | 284 | 250 |
|---|---|---|---|---|

| | | Frozen, sliced,<br>sweetened:<br>10 oz. container | Frozen, sliced<br>sweetened:<br>cup |
|---|---|---|---|
| % | **Water** | 77 | 77 |
| | **Calories** | 250 | 220 |
| grams | **Protein** | 1 | 1 |
| | **Fat** | 3 | 3 |
| | **Carbohydrate** | 64 | 57 |
| | **Calcium** | 11 | 10 |
| milligrams | **Phosphorus** | 37 | 33 |
| | **Iron** | 1.4 | 1.3 |
| | **Sodium** | 6 | 5 |
| | **Potassium** | 352 | 310 |
| | **Thiamin** | .03 | .03 |
| | **Riboflavin** | .11 | .10 |
| | **Niacin** | 2.0 | 1.8 |
| | **Ascorbic Acid** | *116 | *103 |
| unit | **Vitamin A** | 1,850 | 1,630 |

*Value represents products with added ascorbic acid. For products without
added ascorbic acid, value in milligrams is 116 for a 10 oz. container, 103 for 1
cup.

185

# PEARS

| grams | Weight: | 180 | 155 | 200 |
|---|---|---|---|---|
| | | Bartlett, 2½" diam.* | Bosc. 2½" diam.** | D'Anjou 3" diam.* |

| | | | | |
|---|---|---|---|---|
| % | Water | 83 | 83 | 83 |
| grams { | Calories | 100 | 86 | 120 |
| | Protein | 1 | 1 | 1 |
| | Fat | 7 | .5 | 1 |
| | Carbohydrate | 25 | 21.6 | 31 |
| milligrams { | Calcium | 13 | 11 | 16 |
| | Phosphorus | 18 | 16 | 22 |
| | Iron | .5 | .4 | .6 |
| | Sodium | 3 | 3 | 4 |
| | Potassium | 213 | 183 | 260 |
| | Thiamin | .03 | .03 | .04 |
| | Riboflavin | .07 | .06 | .08 |
| | Niacin | .2 | .1 | .2 |
| | Ascorbic Acid | 7 | 6 | 8 |
| unit | Vitamin A | 30 | 30 | 40 |

*About 2½ per lb. with cores and stems.
**About 3 per lb. with cores and stems.

# PEARS
## CANNED

| | grams | | |
|---|---|---|---|
| **Weight:** | | 255 | 244 |

| | | solids & liquid, syrup pack, heavy (halves or sliced) | solids & liquid water packed (halves or sliced) |
|---|---|---|---|
| % | **Water** | 80 | 91.1 |
| | **Calories** | 195 | 78 |
| grams { | **Protein** | 1 | .5 |
| | **Fat** | 1 | .5 |
| | **Carbohydrate** | 50 | 20.3 |
| | **Calcium** | 13 | 12 |
| milligrams { | **Phosphorus** | 18 | 17 |
| | **Iron** | .5 | .5 |
| | **Sodium** | 3 | 2 |
| | **Potassium** | 214 | 215 |
| | **Thiamin** | .03 | .02 |
| | **Riboflavin** | .05 | .05 |
| | **Niacin** | .3 | .2 |
| | **Ascorbic Acid** | 3 | 2 |
| unit | **Vitamin A** | 10 | 10 |

187

# PINEAPPLE

grams **Weight:** 155

Raw, diced

| | | |
|---|---|---|
| % | **Water** | 85 |
| | **Calories** | 80 |
| grams | **Protein** | 1 |
| | **Fat** | Trace |
| | **Carbohydrate** | 21 |
| | **Calcium** | 26 |
| | **Phosphorus** | 12 |
| | **Iron** | .8 |
| | **Sodium** | 2 |
| milligrams | **Potassium** | 226 |
| | **Thiamin** | .14 |
| | **Riboflavin** | .05 |
| | **Niacin** | .3 |
| | **Ascorbic Acid** | 26 |
| unit | **Vitamin A** | 110 |

# PINEAPPLE
## CANNED

| grams | Weight: | 255 | 105 | 58 |
|---|---|---|---|---|
| | |  |  | |
| | | canned,*<br>crushed<br>chunks,<br>tidbits | canned,*<br>slices &<br>liquid; large | canned*<br>slices &<br>liquid<br>medium |
| % | Water | 80 | 80 | 80 |
| | Calories | 190 | 80 | 45 |
| | Protein | 1 | Trace | Trace |
| grams | Fat | Trace | Trace | Trace |
| | Carbohydrate | 49 | 20 | 11 |
| | Calcium | 28 | 12 | 6 |
| | Phosphorus | 13 | 5 | 3 |
| | Iron | .8 | .3 | .2 |
| | Sodium | 3 | 1 | 1 |
| milligrams | Potassium | 245 | 101 | 56 |
| | Thiamin | .20 | .08 | .05 |
| | Riboflavin | .05 | .02 | .01 |
| | Niacin | .5 | .2 | .1 |
| | Ascorbic Acid | 18 | 7 | 4 |
| unit | Vitamin A | 130 | 50 | 30 |

*Heavy syrup pack, solids and liquid.

# JAPANESE
# PLUMS
### PRUNE-TYPE

| grams | **Weight:** | 70 | 30 |
|---|---|---|---|

|  |  | without pits:<br>Japanese &<br>hyrid* | without pits:<br>prune-type** |
|---|---|---|---|
| % | **Water** | 87 | 79 |
| | **Calories** | 32 | 21 |
| grams | **Protein** | .3 | .2 |
| | **Fat** | .1 | .1 |
| | **Carbohydrate** | 8 | 5.6 |
| | **Calcium** | 8 | 3 |
| milligrams | **Phosphorus** | 12 | 5 |
| | **Iron** | .3 | .1 |
| | **Sodium** | 1 | Trace |
| | **Potassium** | 112 | 48 |
| | **Thiamin** | .02 | .01 |
| | **Riboflavin** | .02 | .01 |
| | **Niacin** | .3 | .1 |
| | **Ascorbic Acid** | 4 | 1 |
| unit | **Vitamin A** | 160 | 80 |

*2⅛″ diam., about 6½ per lb. with pits.
**1½″ diam., about 15 per lb. with pits.

# ITALIAN PRUNES
# PLUMS
## CANNED

| grams | Weight: | 272 | 140 |
|---|---|---|---|

|  |  | *heavy syrup pack | **heavy syrup pack |
|---|---|---|---|
| % | **Water** | 77 | 77 |
|  | **Calories** | 215 | 110 |
| grams | **Protein** | 1 | 1 |
|  | **Fat** | Trace | Trace |
|  | **Carbohydrate** | 56 | 29 |
|  | **Calcium** | 23 | 12 |
| milligrams | **Phosphorus** | 26 | 13 |
|  | **Iron** | 2.3 | 1.2 |
|  | **Sodium** | .3 | .1 |
|  | **Potassium** | 367 | 189 |
|  | **Thiamin** | .05 | .03 |
|  | **Riboflavin** | .05 | .03 |
|  | **Niacin** | 1.0 | .5 |
|  | **Ascorbic Acid** | 5 | 3 |
| unit | **Vitamin A** | 3,130 | 1,610 |

*Weight includes pits. After removal of the pits, the weight of the edible portion is 258g for cup, 133g for portion.
**1 portion (3 plums) plus 2¾ tbs. liquid.

# PRUNES
## DRIED

| grams | **Weight:** | 49 | 250 |
|---|---|---|---|

|  |  | 5 uncooked* | cooked, un-sweetened, all sizes, fruit & liquid |
|---|---|---|---|
| % | **Water** | 28 | 66 |
| grams { | **Calories** | 110 | 255 |
|  | **Protein** | 1 | 2 |
|  | **Fat** | Trace | 1 |
|  | **Carbohydrate** | 29 | 67 |
|  | **Calcium** | 22 | 51 |
| milligrams { | **Phosphorus** | 34 | 79 |
|  | **Iron** | 1.7 | 3.8 |
|  | **Sodium** | 12.5 | 9 |
|  | **Potassium** | 298 | 695 |
|  | **Thiamin** | .04 | .07 |
|  | **Riboflavin** | .07 | .15 |
|  | **Niacin** | .7 | 1.5 |
|  | **Ascorbic Acid** | 1 | 2 |
| unit | **Vitamin A** | 690 | 1,590 |

*Weight includes pits. After removal of the pits, the weight of the edible portion is 43g. for uncooked, and 213g for cooked.

# RAISINS

| | cup, not pressed down | packet, ½ oz. (1½ tbsp.) | |
|---|---|---|---|
| grams | **Weight:** 145 | 14 | |

| | | cup, not pressed down | packet, ½ oz. (1½ tbsp.) | |
|---|---|---|---|---|
| % | **Water** | 18 | 18 | |
| | **Calories** | 420 | 40 | |
| grams | **Protein** | 4 | Trace | |
| | **Fat** | Trace | Trace | |
| | **Carbohydrate** | 112 | 11 | |
| | **Calcium** | 90 | 9 | |
| milligrams | **Phosphorus** | 146 | 14 | |
| | **Iron** | 5.1 | 5 | |
| | **Sodium** | 39 | 4 | |
| | **Potassium** | 1,106 | 107 | |
| | **Thiamin** | .16 | .02 | |
| | **Riboflavin** | .12 | .01 | |
| | **Niacin** | .7 | .1 | |
| | **Ascorbic Acid** | 1 | Trace | |
| unit | **Vitamin A** | 30 | Trace | |

# RASPBERRIES
## (RED & BLACK)

| | grams | Weight: | 123 | 134 |
|---|---|---|---|---|

| | | raw, red capped, whole | raw, black |
|---|---|---|---|
| % | Water | 84 | 80.8 |
| grams | Calories | 70 | 98 |
| | Protein | 1 | 2.0 |
| | Fat | 1 | 1.9 |
| | Carbohydrate | 17 | 21.0 |
| | Calcium | 27 | 40 |
| milligrams | Phosphorus | 27 | 29 |
| | Iron | 1.1 | 1.2 |
| | Sodium | 1 | 1 |
| | Potassium | 207 | 267 |
| | Thiamin | .04 | .04 |
| | Riboflavin | .11 | .12 |
| | Niacin | 1.1 | 1.2 |
| | Ascorbic Acid | 31 | 24 |
| unit | Vitamin A | 160 | Trace |

194

# RED
# RASPBERRIES
## FROZEN

grams    **Weight:**         284

sweetened,
10 oz.
container

| | | | |
|---|---|---|---|
| % | **Water** | 74 | |
| | **Calories** | 280 | |
| grams | **Protein** | 2 | |
| | **Fat** | 1 | |
| | **Carbohydrate** | 70 | |
| | **Calcium** | 37 | |
| milligrams | **Phosphorus** | 48 | |
| | **Iron** | 1.7 | |
| | **Sodium** | 3 | |
| | **Potassium** | 284 | |
| | **Thiamin** | .06 | |
| | **Riboflavin** | .17 | |
| | **Niacin** | 1.7 | |
| | **Ascorbic Acid** | 60 | |
| unit | **Vitamin A** | 200 | |

# RHUBARB
## COOKED

| grams | **Weight:** | 270 | 270 |
|---|---|---|---|

| | | from raw diced | from frozen, sweetened |
|---|---|---|---|
| % | **Water** | 63 | 63 |
| | **Calories** | 380 | 385 |
| grams { | **Protein** | 1 | 1 |
| | **Fat** | Trace | 1 |
| | **Carbohydrate** | 97 | 98 |
| | **Calcium** | 211 | 211 |
| milligrams { | **Phosphorus** | 41 | 32 |
| | **Iron** | 1.6 | 1.9 |
| | **Sodium** | 5 | 8 |
| | **Potassium** | 548 | 475 |
| | **Thiamin** | .05 | .05 |
| | **Riboflavin** | .14 | .11 |
| | **Niacin** | .8 | .5 |
| | **Ascorbic Acid** | 16 | 16 |
| unit | **Vitamin A** | 220 | 190 |

# STRAWBERRIES

grams    **Weight:**      149

| | Whole berries, capped |
|---|---|
| %    **Water** | 90 |
| **Calories** | 55 |
| **Protein** | 1 |
| **Fat** | 1 |
| **Carbohydrate** | 13 |
| **Calcium** | 31 |
| **Phosphorus** | 31 |
| **Iron** | 1.5 |
| **Sodium** | 1 |
| **Potassium** | 244 |
| **Thiamin** | .04 |
| **Riboflavin** | .10 |
| **Niacin** | .9 |
| **Ascorbic Acid** | 88 |
| **Vitamin A** | 90 |

grams { (Protein, Fat, Carbohydrate, Calcium)

milligrams { (Phosphorus, Iron, Sodium, Potassium, Thiamin, Riboflavin, Niacin, Ascorbic Acid)

unit (Vitamin A)

# STRAWBERRIES
## FROZEN

| | grams | Weight: | 284 | 454 |
|---|---|---|---|---|

| | | *frozen, sweetened, sliced, 10 oz. container | *frozen, sweetened, 1 lb. container (ab. 1¾ cups) |
|---|---|---|---|
| % | **Water** | 71 | 76 |
| grams { | **Calories** | 310 | 415 |
| | **Protein** | 1 | 2 |
| | **Fat** | 1 | 1 |
| | **Carbohydrate** | 79 | 107 |
| | **Calcium** | 40 | 59 |
| milligrams { | **Phosphorus** | 48 | 73 |
| | **Iron** | 2.0 | 2.7 |
| | **Sodium** | 3 | 5 |
| | **Potassium** | 318 | 472 |
| | **Thiamin** | .06 | .09 |
| | **Riboflavin** | .17 | .27 |
| | **Niacin** | 1.4 | 2.3 |
| | **Ascorbic Acid** | 151 | 249 |
| unit | **Vitamin A** | 90 | 140 |

*Measurement applies to thawed product.

# STRAWBERRIES
## FROZEN

| | grams | **Weight:** | 255 | 255 |
|---|---|---|---|---|

| | | *frozen sweetened, sliced | *frozen sweetened whole |
|---|---|---|---|
| % | **Water** | 71.3 | 75.7 |
| | **Calories** | 278 | 235 |
| grams { | **Protein** | 1.3 | 1.0 |
| | **Fat** | .5 | .5 |
| | **Carbohydrate** | 70.9 | 59.9 |
| | **Calcium** | 36 | 33 |
| milligrams { | **Phosphorus** | 43 | 41 |
| | **Iron** | 1.8 | 1.5 |
| | **Sodium** | 3 | 3 |
| | **Potassium** | 286 | 265 |
| | **Thiamin** | .05 | .05 |
| | **Riboflavin** | .15 | .15 |
| | **Niacin** | 1.3 | 1.3 |
| | **Ascorbic Acid** | 135 | 140 |
| unit | **Vitamin A** | 80 | 80 |

*Measurement applies to thawed product.

# TANGERINE

grams | **Weight:** | 116

( 1 )

*

| | | |
|---|---|---|
| % | **Water** | 87 |
| | **Calories** | 39 |
| grams | **Protein** | .7 |
| | **Fat** | .2 |
| | **Carbohydrate** | 10.0 |
| | **Calcium** | 34 |
| | **Phosphorus** | 15 |
| | **Iron** | .3 |
| | **Sodium** | 2 |
| milligrams | **Potassium** | 108 |
| | **Thiamin** | .05 |
| | **Riboflavin** | .02 |
| | **Niacin** | .1 |
| | **Ascorbic Acid** | 27 |
| unit | **Vitamin A** | 360 |

*2⅜" diam., size 176, without peel (about 4 per lb. with peels and seeds).

# WATERMELON

grams    **Weight:**         926

**

| | | |
|---|---|---|
| % | Water | 93 |
| | Calories | 110 |
| grams | Protein | 2 |
| | Fat | 1 |
| | Carbohydrate | 27 |
| | Calcium | 30 |
| | Phosphorus | 43 |
| milligrams | Iron | 2.1 |
| | Sodium | 4 |
| | Potassium | 426 |
| | Thiamin | .13 |
| | Riboflavin | .13 |
| | Niacin | .9 |
| | Ascorbic Acid | 30 |
| unit | Vitamin A | 2,510 |

*Weight includes rind and seeds. Without rind and seeds, weight of the edible portion is 426g.
**4 by 8 in. wedge with rind and seed (⅟₁₆ of 32⅔ lb. melon, 10 by 16 in).

201

# FRUIT JUICES

APPLE JUICE AND APPLE CIDER
APRICOT NECTAR
CRANBERRY JUICE COCKTAIL
GRAPEFRUIT JUICE, RAW AND CANNED
GRAPEFRUIT JUICE, FROZEN AND DEHYDRATED
GRAPE JUICE
GRAPE DRINK
LEMON JUICE
LEMON CONCENTRATE
LIME JUICE
ORANGE JUICE, RAW AND CANNED
ORANGE JUICE, FROZEN AND DEHYDRATED
ORANGE AND GRAPEFRUIT JUICE
PINEAPPLE JUICE
PRUNE JUICE
TANGERINE JUICE

# APPLE JUICE & APPLE CIDER

grams    **Weight:**        248

Bottled or
canned

| | % | |
|---|---|---|
| **Water** | 88 | |

| | grams | |
|---|---|---|
| **Calories** | 120 | |
| **Protein** | Trace | |
| **Fat** | Trace | |
| **Carbohydrate** | 30 | |
| **Calcium** | 15 | |

| | milligrams | |
|---|---|---|
| **Phosphorus** | 22 | |
| **Iron** | 1.5 | |
| **Sodium** | 2 | |
| **Potassium** | 250 | |
| **Thiamin** | .02 | |
| **Riboflavin** | .05 | |
| **Niacin** | .2 | |
| **Ascorbic Acid** | *2 | |

| | unit | |
|---|---|---|
| **Vitamin A** | N.A. | |

*Applies to product without added ascorbic acid. For value of product with added ascorbic acid, refer to label.

# APRICOT NECTAR

| | grams | **Weight:** | 251 |

canned

| | | | |
|---|---|---|---|
| % | **Water** | 85 | |
| | **Calories** | 145 | |
| grams | **Protein** | 1 | |
| | **Fat** | Trace | |
| | **Carbohydrate** | 37 | |
| | **Calcium** | 23 | |
| | **Phosphorus** | 30 | |
| | **Iron** | .5 | |
| milligrams | **Sodium** | Trace | |
| | **Potassium** | 379 | |
| | **Thiamin** | .03 | |
| | **Riboflavin** | .03 | |
| | **Niacin** | .5 | |
| | **Ascorbic Acid** | *36 | |
| unit | **Vitamin A** | 2,380 | |

*Based on product with label claim of 45% of U.S. RDA in 6 fl. oz.

# CRANBERRY JUICE
## COCKTAIL

| | | |
|---|---|---|
| grams | **Weight:** | 253 |

sweetened

| | | |
|---|---|---|
| % | **Water** | 83 |
| | **Calories** | 165 |
| grams | **Protein** | Trace |
| | **Fat** | Trace |
| | **Carbohydrate** | 42 |
| | **Calcium** | 13 |
| milligrams | **Phosphorus** | 8 |
| | **Iron** | 8 |
| | **Sodium** | 3 |
| | **Potassium** | 25 |
| | **Thiamin** | .03 |
| | **Riboflavin** | .03 |
| | **Niacin** | .1 |
| | **Ascorbic Acid** | *81 |
| unit | **Vitamin A** | Trace |

*Based on product with label claim of 100% of U.S. RDA in 6 fl. oz.

205

# GRAPEFRUIT JUICE

| grams | Weight: | 246 | 247 | 250 |
|---|---|---|---|---|
| | | 1 | 1 | 1 |
| | | Raw, pink, red or white | canned, white, un-sweetened | canned, white, sweetened |
| % | Water | 90 | 89 | 86 |
| grams { | Calories | 95 | 100 | 135 |
| | Protein | 1 | 1 | 1 |
| | Fat | Trace | Trace | Trace |
| | Carbohydrate | 23 | 24 | 32 |
| milligrams { | Calcium | 22 | 20 | 20 |
| | Phosphorus | 37 | 35 | 35 |
| | Iron | .5 | 1.0 | 1.0 |
| | Sodium | 2 | 2 | 3 |
| | Potassium | 399 | 400 | 405 |
| | Thiamin | .10 | .07 | .08 |
| | Riboflavin | .05 | .05 | .05 |
| | Niacin | .5 | .5 | .5 |
| | Ascorbic Acid | 93 | 84 | 78 |
| unit | Vitamin A | * | 20 | 30 |

*For white-fleshed varieties, value is about 20 I.U. per cup; for red-fleshed varieties, 1,080 I.U.

# GRAPEFRUIT JUICE
## FROZEN AND DEHYDRATED

| grams | **Weight:** | 207 | 247 | 247 |
|---|---|---|---|---|
| | |  frozen, concentrate unsweetened: undiluted, 6 fl. oz. can | frozen concentrate: unsweetened, diluted with 3 parts water by vol. | dehydrated crystals, prepared with water |
| % | **Water** | 62 | 89 | 90 |
| | **Calories** | 300 | 100 | 100 |
| grams | **Protein** | 4 | 1 | 1 |
| | **Fat** | 1 | Trace | Trace |
| | **Carbohydrate** | 72 | 24 | 24 |
| | **Calcium** | 70 | 25 | 22 |
| milligrams | **Phosphorus** | 124 | 42 | 40 |
| | **Iron** | .8 | .2 | .2 |
| | **Sodium** | 8 | 2 | 2 |
| | **Potassium** | 1,250 | 420 | 412 |
| | **Thiamin** | .29 | .10 | .10 |
| | **Riboflavin** | .12 | .04 | .05 |
| | **Niacin** | 1.4 | .5 | .5 |
| | **Ascorbic Acid** | 286 | 96 | 91 |
| unit | **Vitamin A** | 60 | 20 | 20 |

# GRAPE JUICE

| grams | Weight: | 253 | 216 | 250 |
|---|---|---|---|---|
| | |  canned or bottled |  frozen con-centrate, sweet-ened: undiluted, 6 fl. oz. can |  frozen con-centrate, sweet-ened: diluted with 3 parts water by vol. |
| % | Water | 83 | 53 | 86 |
| | Calories | 165 | 395 | 135 |
| grams { | Protein | 1 | 1 | 1 |
| | Fat | Trace | Trace | Trace |
| | Carbohydrate | 42 | 100 | 33 |
| milligrams { | Calcium | 28 | 22 | 8 |
| | Phosphorus | 30 | 32 | 10 |
| | Iron | .8 | .9 | .3 |
| | Sodium | 5 | 6 | 3 |
| | Potassium | 293 | 255 | 85 |
| | Thiamin | .10 | .13 | .05 |
| | Riboflavin | .05 | .22 | .08 |
| | Niacin | .5 | 1.5 | .5 |
| | Ascorbic Acid | *Trace | **32 | **10 |
| unit | Vitamin A | N.A. | 40 | 10 |

*Applies to product without added ascorbic acid. See label. ** With added ascorbic acid, based on claim that 6 fl. oz. of reconstituted juice contain 45% or 50% of the U.S. RDA, value in | milligrams is 108 or 120 for a 6 fl. oz can, 36 or 40 for 1 cup of diluted juice.

# GRAPE DRINK

| grams | **Weight:** | 250 |

canned

| | | |
|---|---|---|
| % | **Water** | 86 |
| | **Calories** | 135 |
| grams | **Protein** | Trace |
| | **Fat** | Trace |
| | **Carbohydrate** | 35 |
| | **Calcium** | 8 |
| milligrams | **Phosphorus** | 10 |
| | **Iron** | .3 |
| | **Sodium** | 3 |
| | **Potassium** | 88 |
| | **Thiamin** | *.03 |
| | **Riboflavin** | *.3 |
| | **Niacin** | .3 |
| | **Ascorbic Acid** | * |
| unit | **Vitamin A** | N.A. |

*For products with added thiamin and riboflavin but without added ascorbic acid, values in milligrams would be 0.60 for thiamin, 0.80 for riboflavin, and trace for ascorbic acid. For products with only ascorbic acid added, value varies with the brand. Consult the label.

209

# LEMON JUICE

| grams | **Weight:** | 244 | 244 | 183 |
|---|---|---|---|---|
| | |  | |  |
| | | raw | canned, or bottled, unsweetened | frozen, single strength, unsweetened, 6 fl. oz. can |
| % | **Water** | 91 | 92 | 92 |
| | **Calories** | 60 | 55 | 40 |
| grams | **Protein** | 1 | 1 | 1 |
| | **Fat** | Trace | Trace | Trace |
| | **Carbohydrate** | 20 | 19 | 13 |
| | **Calcium** | 17 | 17 | 13 |
| milligrams | **Phosphorus** | 24 | 24 | 16 |
| | **Iron** | .5 | .5 | .5 |
| | **Sodium** | 2 | 2 | 2 |
| | **Potassium** | 344 | 344 | 258 |
| | **Thiamin** | .07 | .07 | .05 |
| | **Riboflavin** | .02 | .02 | .02 |
| | **Niacin** | .2 | .2 | .2 |
| | **Ascorbic Acid** | 112 | 102 | 81 |
| unit | **Vitamin A** | 50 | 50 | 40 |

# LEMON
## FROZEN, CONCENTRATE

| | | |
|---|---|---|
| grams | **Weight:** | 219 | 248 |

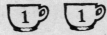

| | | undiluted<br>6 fl. oz. can | diluted with<br>4⅓ parts<br>water by vol. |
|---|---|---|---|
| % | **Water** | 49 | 89 |
| | **Calories** | 425 | 105 |
| grams | **Protein** | Trace | Trace |
| | **Fat** | Trace | Trace |
| | **Carbohydrate** | 112 | 28 |
| | **Calcium** | 9 | 2 |
| milligrams | **Phosphorus** | 13 | 3 |
| | **Iron** | .4 | .1 |
| | **Sodium** | 4 | 1 |
| | **Potassium** | 153 | 40 |
| | **Thiamin** | .05 | .01 |
| | **Riboflavin** | .06 | .02 |
| | **Niacin** | .7 | .2 |
| | **Ascorbic Acid** | 66 | 17 |
| unit | **Vitamin A** | 40 | 10 |

211

# LIME JUICE

| grams | **Weight:** | 246 | 246 |
|---|---|---|---|

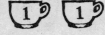

| | | raw | canned, unsweetened |
|---|---|---|---|
| % | **Water** | 90 | 90 |
| grams | **Calories** | 65 | 65 |
| | **Protein** | 1 | 1 |
| | **Fat** | Trace | Trace |
| | **Carbohydrate** | 22 | 22 |
| | **Calcium** | 22 | 22 |
| milligrams | **Phosphorus** | 27 | 27 |
| | **Iron** | .5 | .5 |
| | **Sodium** | 2 | 2 |
| | **Potassium** | 256 | 256 |
| | **Thiamin** | .05 | .05 |
| | **Riboflavin** | .02 | .02 |
| | **Niacin** | .2 | .2 |
| | **Ascorbic Acid** | 79 | 52 |
| unit | **Vitamin A** | 20 | 20 |

# ORANGE JUICE

| | grams | **Weight:** | 248 | 249 |
|---|---|---|---|---|

| | raw<br>all varieties | canned,<br>unsweetened |
|---|---|---|
| **Water** % | 88 | 87 |
| **Calories** | 110 | 120 |
| **Protein** | 2 | 2 |
| **Fat** | Trace | Trace |
| **Carbohydrate** | 26 | 28 |
| **Calcium** | 27 | 25 |
| **Phosphorus** | 42 | 45 |
| **Iron** | .5 | 1.0 |
| **Sodium** | 2 | 2 |
| **Potassium** | 496 | 496 |
| **Thiamin** | .22 | .17 |
| **Riboflavin** | .07 | .05 |
| **Niacin** | 1.0 | .7 |
| **Ascorbic Acid** | 124 | 100 |
| **Vitamin A** | 500 | 500 |

grams / milligrams / unit

# ORANGE JUICE
## FROZEN AND DEHYDRATED

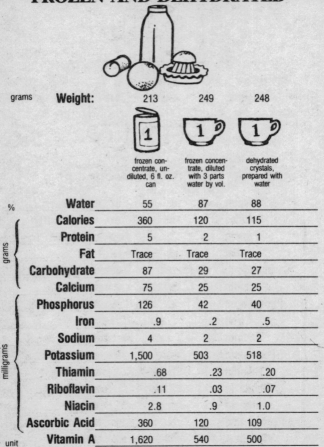

| grams | Weight: | 213 | 249 | 248 |
|---|---|---|---|---|
| | | frozen concentrate, undiluted, 6 fl. oz. can | frozen concentrate, diluted with 3 parts water by vol. | dehydrated crystals, prepared with water |
| % | **Water** | 55 | 87 | 88 |
| | **Calories** | 360 | 120 | 115 |
| grams | **Protein** | 5 | 2 | 1 |
| | **Fat** | Trace | Trace | Trace |
| | **Carbohydrate** | 87 | 29 | 27 |
| | **Calcium** | 75 | 25 | 25 |
| milligrams | **Phosphorus** | 126 | 42 | 40 |
| | **Iron** | .9 | .2 | .5 |
| | **Sodium** | 4 | 2 | 2 |
| | **Potassium** | 1,500 | 503 | 518 |
| | **Thiamin** | .68 | .23 | .20 |
| | **Riboflavin** | .11 | .03 | .07 |
| | **Niacin** | 2.8 | .9 | 1.0 |
| | **Ascorbic Acid** | 360 | 120 | 109 |
| unit | **Vitamin A** | 1,620 | 540 | 500 |

# ORANGE & GRAPEFRUIT JUICE
## FROZEN CONCENTRATE

| grams | Weight: | 210 | 248 |
|---|---|---|---|

| | | undiluted, 6 fl. oz. can | diluted with 3 pts. water by volume |
|---|---|---|---|
| % | Water | 59 | 88 |
| | Calories | 330 | 110 |
| grams | Protein | 4 | 1 |
| | Fat | 1 | Trace |
| | Carbohydrate | 78 | 26 |
| | Calcium | 61 | 20 |
| milligrams | Phosphorus | 99 | 32 |
| | Iron | .8 | .2 |
| | Sodium | N.A. | 1 |
| | Potassium | 1,308 | 439 |
| | Thiamin | .48 | .15 |
| | Riboflavin | .06 | .02 |
| | Niacin | 2.3 | .7 |
| | Ascorbic Acid | 302 | 102 |
| unit | Vitamin A | 800 | 270 |

215

# PINEAPPLE JUICE
## CANNED

grams  **Weight:**  250

unsweetend

| | | |
|---|---|---|
| % | **Water** | 86 |
| | **Calories** | 140 |
| grams | **Protein** | 1 |
| | **Fat** | Trace |
| | **Carbohydrate** | 34 |
| | **Calcium** | 38 |
| milligrams | **Phosphorus** | 23 |
| | **Iron** | .8 |
| | **Sodium** | 3 |
| | **Potassium** | 373 |
| | **Thiamin** | .13 |
| | **Riboflavin** | .05 |
| | **Niacin** | .5 |
| | **Ascorbic Acid** | *80 |
| unit | **Vitamin A** | 130 |

*Based on product with label claim of 100% of U.S. RDA in 6 fl. oz.

# PRUNE JUICE

grams **Weight:** 256

canned or
bottled

| | | |
|---|---|---|
| % | **Water** | 80 |
| | **Calories** | 195 |
| grams { | **Protein** | 1 |
| | **Fat** | Trace |
| | **Carbohydrate** | 49 |
| | **Calcium** | 36 |
| milligrams { | **Phosphorus** | 51 |
| | **Iron** | 1.8 |
| | **Sodium** | 5 |
| | **Potassium** | 602 |
| | **Thiamin** | .03 |
| | **Riboflavin** | .03 |
| | **Niacin** | 1.0 |
| | **Ascorbic Acid** | 5 |
| unit | **Vitamin A** | N.A. |

217

# TANGERINE JUICE
## (CONCENTRATE)

| | | sweetened canned | unsweetened canned | unsweetened frozen* |
|---|---|---|---|---|
| grams | **Weight:** | 249 | 247 | 248 |
| % | **Water** | 87 | 88.8 | 88.1 |
| grams | **Calories** | 125 | 106 | 114 |
| | **Protein** | 1 | 1.2 | 1.2 |
| | **Fat** | Trace | .5 | .5 |
| | **Carbohydrate** | 30 | 25.2 | 26.8 |
| | **Calcium** | 44 | 44 | 45 |
| milligrams | **Phosphorus** | 35 | 35 | 35 |
| | **Iron** | .5 | .5 | .5 |
| | **Sodium** | 2 | 2 | 2 |
| | **Potassium** | 440 | 440 | 432 |
| | **Thiamin** | .15 | .15 | .14 |
| | **Riboflavin** | .05 | .05 | .04 |
| | **Niacin** | .2 | .2 | .3 |
| | **Ascorbic Acid** | 54 | 54 | 67 |
| unit | **Vitamin A** | 1,040 | 1,040 | 1,020 |

*Diluted with 3 parts water by volume.

# VEGETABLES

ARTICHOKES
ASPARAGUS
BAMBOO SHOOTS
BEANS, GREEN
BEANS, LIMA
BEANS, MUNG, SPROUTS
BEANS, YELLOW
BEET GREENS
BEETS
BLACKEYE PEAS
BROCCOLI
BRUSSELS SPROUTS
CABBAGE, (WONGBOK OR PE-TSAI)
CABBAGE, GREEN (COMMON)
CABBAGE, RED
CABBAGE, SAVOY
CABBAGE, WHITE MUSTARD
(BOKCHOY OR PAKCHOY)
CARROTS
CAULIFLOWER
CELERY, PASCAL TYPE
CHARD, SWISS
CHICORY, WITLOOF
CHIVES
COLLARDS

CORN, COOKED
CORN, CANNED
CUCUMBER PICKLES
CUCUMBER SLICES
DANDELION GREENS
ENDIVE
KALE
LETTUCE, BUTTERHEAD VARIETIES
LETTUCE, CRISPHEAD VARIETIES
LETTUCE, LOOSELEAF VARIETIES
MIXED VEGETABLES
MUSHROOMS
MUSTARD GREENS
OKRA PODS
OLIVES
ONIONS, GREEN
ONIONS, MATURE
PARSLEY
PARSNIPS
PEAS, GREEN
PEPPERS, HOT
PEPPERS, SWEET
POTATO CHIPS
POTATO SALAD
POTATOES, BAKED
POTATOES, BOILED
POTATOES, FRENCH FRIED
POTATOES, HASHED BROWN

POTATOES, MASHED
PUMPKIN
RADISHES
SAUERKRAUT
SQUASH, SUMMER VARIETIES
SQUASH, WINTER VARIETIES
SPINACH
SWEET POTATOES
TOMATOES
TOMATO JUICE
TURNIP GREENS
TURNIPS

# ARTICHOKES
## GLOBE OR FRENCH

| | grams | Weight: | 380 | 250 |
|---|---|---|---|---|

| | | 1 | 1 |
|---|---|---|---|
| | | cooked*<br>(24 per<br>½ box) | cooked*<br>(36 per<br>½ box) |
| % | Water | 86.5 | 86.5 |
| | Calories | ** | *** |
| grams { | Protein | 4.3 | 2.8 |
| | Fat | .3 | .2 |
| | Carbohydrate | 15 | 9.9 |
| | Calcium | 78 | 51 |
| milligrams { | Phosphorus | 105 | 69 |
| | Iron | 1.7 | 1.1 |
| | Sodium | ****46 | ****30 |
| | Potassium | 458 | 301 |
| | Thiamin | .11 | .07 |
| | Riboflavin | .06 | .04 |
| | Niacin | 1.1 | .7 |
| | Ascorbic Acid | 12 | 8 |
| unit | Vitamin A | 230 | 150 |

*Refuse: stem and inedible parts of bracts and flower, 60% — measures and weight include refuse. **Varies from 12 cal. fresh to 67 cal. stored.
***Varies from 8 cal. fresh to 44 cal. stored.
****Value for unsalted product. If salted, add 236 mg. per 100 g. of vegetable.

# GREEN
# ASPARAGUS
## CUTS AND TIPS

| | grams | **Weight:** | 145 | 180 |
|---|---|---|---|---|

| | | cuts & tips*<br>cooked<br>from raw | cuts & tips*<br>cooked<br>from frozen |
|---|---|---|---|
| % | **Water** | 94 | 93 |
| | **Calories** | 30 | 40 |
| | **Protein** | 3 | 6 |
| grams | **Fat** | Trace | Trace |
| | **Carbohydrate** | 5 | 6 |
| | **Calcium** | 30 | 40 |
| | **Phosphorus** | 73 | 115 |
| | **Iron** | 0.9 | 2.2 |
| | **Sodium** | 3 | 2 |
| milligrams | **Potassium** | 265 | 396 |
| | **Thiamin** | .23 | .25 |
| | **Riboflavin** | .26 | .23 |
| | **Niacin** | 2.0 | 1.8 |
| | **Ascorbic Acid** | 38 | 41 |
| unit | **Vitamin A** | 1,310 | 1,530 |

*1½ to 2 in. lengths

# GREEN
# ASPARAGUS
### SPEARS

| | grams | Weight: | 60 | 60 | 60 |
|---|---|---|---|---|---|
| | | | spears*<br>cooked<br>from raw | spears*<br>cooked<br>from frozen | spears*<br>canned |
| % | | **Water** | 94 | 92 | 93 |
| | | **Calories** | 10 | 15 | 15 |
| grams | | **Protein** | 1 | 2 | 2 |
| | | **Fat** | Trace | Trace | Trace |
| | | **Carbohydrate** | 2 | 2 | 3 |
| | | **Calcium** | 13 | 13 | 15 |
| milligrams | | **Phosphorus** | 30 | 40 | 42 |
| | | **Iron** | .4 | .7 | 1.5 |
| | | **Sodium** | **1 | **1 | **189 |
| | | **Potassium** | 110 | 143 | 133 |
| | | **Thiamin** | .10 | .10 | .05 |
| | | **Riboflavin** | .11 | .08 | .08 |
| | | **Niacin** | .8 | .7 | .6 |
| | | **Ascorbic Acid** | 16 | 16 | 12 |
| unit | | **Vitamin A** | 540 | 470 | 640 |

*½ in. diam. at base
**Estimated value based on addition of salt in amount of 0.6% of finished product.

# **BAMBOO SHOOTS**

| | | |
|---|---|---|
| grams | **Weight:** | 454 |

raw, cut into
1″ lengths

| % | **Water** | 91.0 |
|---|---|---|
| | **Calories** | 122 |
| grams | **Protein** | 11.8 |
| | **Fat** | 1.4 |
| | **Carbohydrate** | 23.6 |
| | **Calcium** | 59 |
| milligrams | **Phosphorus** | 268 |
| | **Iron** | 2.3 |
| | **Sodium** | N.A. |
| | **Potassium** | 2,418 |
| | **Thiamin** | .68 |
| | **Riboflavin** | .32 |
| | **Niacin** | 2.7 |
| | **Ascorbic Acid** | 18 |
| unit | **Vitamin A** | 90 |

# GREEN SNAP
# BEANS

grams **Weight:** 125

cooked,
drained: from
raw (cuts &
French style)

| | | |
|---|---|---|
| % | **Water** | 92 |
| | **Calories** | 30 |
| grams { | **Protein** | 2 |
| | **Fat** | Trace |
| | **Carbohydrate** | 7 |
| | **Calcium** | 63 |
| milligrams { | **Phosphorus** | 46 |
| | **Iron** | .8 |
| | **Sodium** | *5 |
| | **Potassium** | 189 |
| | **Thiamin** | .09 |
| | **Riboflavin** | .11 |
| | **Niacin** | .6 |
| | **Ascorbic Acid** | 15 |
| unit | **Vitamin A** | 680 |

*Value is for unsalted product. If salt is used, increased value by 236 mg. per
100g of vegetable — an estimated figure based on typical amount of salt (0.6%)
in canned vegetables.

226

# GREEN SNAP
# BEANS
## CANNED, FROZEN

| | grams | Weight: | 135 | 130 | 135 |
|---|---|---|---|---|---|

| | | from frozen: cut, cooked, drained | from frozen: cooked, drained French style | canned, drained solids (cuts) |
|---|---|---|---|---|
| % | Water | 92 | 92 | 92 |
| grams | Calories | 35 | 35 | 30 |
| | Protein | 2 | 2 | 2 |
| | Fat | Trace | Trace | Trace |
| | Carbohydrate | 8 | 8 | 7 |
| milligrams | Calcium | 54 | 49 | 61 |
| | Phosphorus | 43 | 39 | 34 |
| | Iron | .9 | 1.2 | 2.0 |
| | Sodium | *1 | *3 | *319 |
| | Potassium | 205 | 177 | 138 |
| | Thiamin | .09 | .08 | .04 |
| | Riboflavin | .12 | .10 | .07 |
| | Niacin | .5 | .4 | .4 |
| | Ascorbic Acid | 7 | 9 | 5 |
| unit | Vitamin A | 780 | 690 | 630 |

*Value is for unsalted product. If salt is used, increase value by 236 mg. per 100g of vegetable — an estimated figure based on typical amount of salt (0.6%) in canned vegetables.

# LIMA BEANS
## FROZEN, COOKED

| | grams | **Weight:** | 170 | 180 |
|---|---|---|---|---|

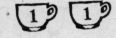

| | | drained: thick-seeded types (Fordhook) | drained: thin-seeded types (Baby limas) |
|---|---|---|---|
| % | **Water** | 74 | 69 |
| | **Calories** | 170 | 210 |
| | **Protein** | 10 | 13 |
| grams | **Fat** | Trace | Trace |
| | **Carbohydrate** | 32 | 40 |
| | **Calcium** | 34 | 63 |
| | **Phosphorus** | 153 | 227 |
| | **Iron** | 2.9 | 4.7 |
| | **Sodium** | *172 | *232 |
| milligrams | **Potassium** | 724 | 709 |
| | **Thiamin** | .12 | .16 |
| | **Riboflavin** | .09 | .09 |
| | **Niacin** | 1.7 | 2.2 |
| | **Ascorbic Acid** | 29 | 2 |
| unit | **Vitamin A** | 390 | 400 |

*Value based on average weighted in accordance with commercial practice in freezing vegetables. Wide range in sodium content occurs. For cooked vegetables, values also represent no additional salting.

# BEAN SPROUTS
## MUNG

| grams | Weight: | 105 | 125 |
|---|---|---|---|

|  |  | raw | cooked, drained |
|---|---|---|---|
| % | **Water** | 89 | 91 |
|  | **Calories** | 35 | 35 |
| grams { | **Protein** | 4 | 4 |
|  | **Fat** | Trace | Trace |
|  | **Carbohydrate** | 7 | 7 |
|  | **Calcium** | 20 | 21 |
| milligrams { | **Phosphorus** | 67 | 60 |
|  | **Iron** | 1.4 | 1.1 |
|  | **Sodium** | 5 | *5 |
|  | **Potassium** | 234 | 195 |
|  | **Thiamin** | .14 | .11 |
|  | **Riboflavin** | .14 | .13 |
|  | **Niacin** | .8 | .9 |
|  | **Ascorbic Acid** | 20 | 8 |
| unit | **Vitamin A** | 20 | 30 |

*Value is for unsalted product. If salt is used, increase value by 236 mg. per 100g. of vegetable — an estimated figure based on typical amount of salt (0.6%) in canned vegetables.

229

# BEANS
## YELLOW OR WAX SNAP

| | grams | Weight: | 125 | 135 | 135 |
|---|---|---|---|---|---|

| | | cooked, drained: from raw (cuts & French style) | cooked, drained: from frozen (cuts) | canned, drained solids (cuts) |
|---|---|---|---|---|
| % | Water | 93 | 92 | 92 |
| | Calories | 30 | 35 | 30 |
| grams | Protein | 2 | 2 | 2 |
| | Fat | Trace | Trace | Trace |
| | Carbohydrate | 6 | 8 | 7 |
| | Calcium | 63 | 47 | 61 |
| milligrams | Phosphorus | 46 | 42 | 34 |
| | Iron | .8 | .9 | 2.0 |
| | Sodium | *4 | *1 | *3 |
| | Potassium | 189 | 221 | 128 |
| | Thiamin | .09 | .09 | .04 |
| | Riboflavin | .11 | .11 | .07 |
| | Niacin | .6 | .5 | .4 |
| | Ascorbic Acid | 16 | 8 | 7 |
| unit | Vitamin A | 290 | 140 | 140 |

*Value is for unsalted product. If salt is used, increase value by 236 mg. per 100g of vegetable — an estimated figure based on typical amount of salt (0.6%) in canned vegetables.

# BEET GREENS

| | | |
|---|---|---|
| grams | **Weight:** | 145 |

leaves &
stems cooked,
drained

| | | |
|---|---|---|
| % | **Water** | 94 |
| | **Calories** | 25 |
| | **Protein** | 2 |
| grams | **Fat** | Trace |
| | **Carbohydrate** | 5 |
| | **Calcium** | 144 |
| | **Phosphorus** | 36 |
| | **Iron** | 2.8 |
| | **Sodium** | *110 |
| milligrams | **Potassium** | 481 |
| | **Thiamin** | .10 |
| | **Riboflavin** | .22 |
| | **Niacin** | .4 |
| | **Ascorbic Acid** | 22 |
| unit | **Vitamin A** | 7,400 |

*Value is for unsalted product. If salt is used, increase value by 236 mg. per
100g of vegetable — an estimated figure based on typical amount of salt (0.6%)
in canned vegetables.

# BEETS

| grams | Weight: | 100 | 170 |
|---|---|---|---|

| | | cooked, drained, peeled: whole 2" diam. | cooked, drained, peeled: diced or sliced |
|---|---|---|---|
| % | Water | 91 | 91 |
| grams { | Calories | 30 | 55 |
| | Protein | 1 | 2 |
| | Fat | Trace | Trace |
| | Carbohydrate | 7 | 12 |
| milligrams { | Calcium | 14 | 24 |
| | Phosphorus | 23 | 39 |
| | Iron | .5 | .9 |
| | Sodium | *43 | *73 |
| | Potassium | 208 | 354 |
| | Thiamin | .03 | .05 |
| | Riboflavin | .04 | .07 |
| | Niacin | .3 | .5 |
| | Ascorbic Acid | 6 | 10 |
| unit | Vitamin A | 20 | 30 |

*Value is for unsalted product. If salt is used, increase value by 236 mg. per 100g of vegetable — an estimated figure based on typical amount of salt (0.6%) in canned vegetables.

# BEETS
## CANNED

| | grams | Weight: | 160 | 170 |
|---|---|---|---|---|

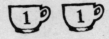

| | | | drained, solids & whole, small | drained, solids: diced or sliced |
|---|---|---|---|---|
| % | | Water | 89 | 89 |
| | | Calories | 60 | 65 |
| | | Protein | 2 | 2 |
| grams | | Fat | Trace | Trace |
| | | Carbohydrate | 14 | 15 |
| | | Calcium | 30 | 32 |
| | | Phosphorus | 29 | 31 |
| | | Iron | 1.1 | 1.2 |
| | | Sodium | *378 | *401 |
| milligrams | | Potassium | 267 | 284 |
| | | Thiamin | .02 | .02 |
| | | Riboflavin | .05 | .05 |
| | | Niacin | .2 | .2 |
| | | Ascorbic Acid | 5 | 5 |
| unit | | Vitamin A | 30 | 30 |

*Estimated value based on addition of salt in amount of 0.6% of finished product.

# BLACKEYE PEAS

| | grams | Weight: | 165 | 170 |
|---|---|---|---|---|

|  |  | cooked, drained from raw | cooked, drained from frozen |
|---|---|---|---|
| % | Water | 72 | 66 |
| grams | Calories | 180 | 220 |
| | Protein | 13 | 15 |
| | Fat | 1 | 1 |
| | Carbohydrate | 30 | 40 |
| | Calcium | 40 | 43 |
| milligrams | Phosphorus | 241 | 286 |
| | Iron | 3.5 | 4.8 |
| | Sodium | *2 | *66 |
| | Potassium | 625 | 573 |
| | Thiamin | .50 | .68 |
| | Riboflavin | .18 | .19 |
| | Niacin | 2.3 | 2.4 |
| | Ascorbic Acid | 28 | 15 |
| unit | Vitamin A | 580 | 290 |

*Value is for unsalted product. If salt is used, increase value by 236 mg. per 100g. of vegetable — an estimated figure based on typical amount of salt (0.6%) in canned vegetables.

# BROCCOLI

| | grams | **Weight:** | 180 | 155 |
|---|---|---|---|---|

| | | cooked, drained from raw stalk, medium size | cooked, drained from raw stalks cut into ½" pieces |
|---|---|---|---|
| % | **Water** | 91 | 91 |
| | **Calories** | 45 | 40 |
| grams | **Protein** | 6 | 5 |
| | **Fat** | 1 | Trace |
| | **Carbohydrate** | 8 | 7 |
| | **Calcium** | 158 | 136 |
| milligrams | **Phosphorus** | 112 | 96 |
| | **Iron** | 1.4 | 1.2 |
| | **Sodium** | *18 | *16 |
| | **Potassium** | 481 | 414 |
| | **Thiamin** | .16 | .14 |
| | **Riboflavin** | .36 | .31 |
| | **Niacin** | 1.4 | 1.2 |
| | **Ascorbic Acid** | 162 | 140 |
| unit | **Vitamin A** | 4,500 | 3,880 |

*Value is for unsalted products. If salt is added, increase value by 236 mg. per 100g of vegetable — an estimated figure based on typical amount of salt (0.6%) in canned vegetables.

# BROCCOLI
## (FROZEN)

grams **Weight:**  30  185

|  | cooked, drained, stalk 4½ to 5" long | cooked, drained, chopped |
|---|---|---|
| **Water** % | 91 | 92 |
| **Calories** | 10 | 50 |
| **Protein** | 1 | 5 |
| **Fat** | Trace | 1 |
| **Carbohydrate** | 1 | 9 |
| **Calcium** | 12 | 100 |
| **Phosphorus** | 17 | 104 |
| **Iron** | .2 | 1.3 |
| **Sodium** | *4 | *28 |
| **Potassium** | 66 | 392 |
| **Thiamin** | .02 | .11 |
| **Riboflavin** | .03 | .22 |
| **Niacin** | .2 | .9 |
| **Ascorbic Acid** | 22 | 105 |
| **Vitamin A** | 570 | 4,810 |

grams { (Water through Calcium)

milligrams { (Phosphorus through Ascorbic Acid)

unit (Vitamin A)

*Value is for unsalted products. If salt is added, increase value by 236 mg. per 100g of vegetable — an estimated figure based on typical amount of salt (0.6%) in canned vegetables.

# BRUSSELS SPROUTS

| | | cooked, drained from raw 7-8 sprouts (1¼ to 1½" diam.) | cooked, drained from frozen |
|---|---|:---:|:---:|
| grams | **Weight:** | 155 | 155 |
| % | **Water** | 88 | 89 |
| | **Calories** | 55 | 50 |
| grams | **Protein** | 7 | 5 |
| | **Fat** | 1 | Trace |
| | **Carbohydrate** | 10 | 10 |
| | **Calcium** | 50 | 33 |
| milligrams | **Phosphorus** | 112 | 95 |
| | **Iron** | 1.7 | 1.2 |
| | **Sodium** | *16 | *22 |
| | **Potassium** | 423 | 457 |
| | **Thiamin** | .12 | .12 |
| | **Riboflavin** | .22 | .16 |
| | **Niacin** | 1.2 | .9 |
| | **Ascorbic Acid** | 135 | 126 |
| unit | **Vitamin A** | 810 | 880 |

*Value is for unsalted product. If salt is used, increase value by 236 mg. per 100g of vegetable — an estimated figure based on typical amount of salt (0.6%) in canned vegetables.

237

# CABBAGE, CHINESE
## (WONGBOK OR PE-TSAI)

grams **Weight:** 75

raw, 1 in.
pieces

| | | |
|---|---|---|
| % | **Water** | 95 |
| | **Calories** | 10 |
| grams | **Protein** | 1 |
| | **Fat** | Trace |
| | **Carbohydrate** | 2 |
| | **Calcium** | 32 |
| | **Phosphorus** | 30 |
| | **Iron** | .5 |
| | **Sodium** | 17 |
| milligrams | **Potassium** | 190 |
| | **Thiamin** | .04 |
| | **Riboflavin** | .03 |
| | **Niacin** | .5 |
| | **Ascorbic Acid** | 19 |
| unit | **Vitamin A** | *110 |

*Value does not allow for losses that might occur from cutting, chopping, or shredding.

# GREEN
# CABBAGE
## (COMMON VARIETIES)

| | Weight: | 70 | 90 | 145 |
|---|---|---|---|---|
| grams | | | | |

| | | raw: coarsely shredded or sliced | raw: finely shredded or chopped | cooked, drained |
|---|---|---|---|---|
| % | **Water** | 92 | 92 | 94 |
| | **Calories** | 15 | 20 | 30 |
| grams | **Protein** | 1 | 1 | 2 |
| | **Fat** | Trace | Trace | Trace |
| | **Carbohydrate** | 4 | 5 | 6 |
| | **Calcium** | 34 | 44 | 64 |
| milligrams | **Phosphorus** | 20 | 26 | 29 |
| | **Iron** | .3 | .4 | .4 |
| | **Sodium** | 14 | 18 | *20 |
| | **Potassium** | 163 | 210 | 236 |
| | **Thiamin** | .04 | .05 | .06 |
| | **Riboflavin** | .04 | .05 | .06 |
| | **Niacin** | .02 | .3 | .4 |
| | **Ascorbic Acid** | 33 | 42 | 48 |
| unit | **Vitamin A** | 90 | 120 | 190 |

*Value is for unsalted product. If salt is used, increase value by 236 mg. per 100g of vegetable — an estimated figure based on typical amount of salt (0.6%) in canned vegetables.

239

# RED
# CABBAGE

grams **Weight:** 70

raw, coarsely
shredded or
sliced

| | | |
|---|---|---|
| % | **Water** | 90 |
| | **Calories** | 20 |
| grams { | **Protein** | 1 |
| | **Fat** | Trace |
| | **Carbohydrate** | 5 |
| | **Calcium** | 29 |
| milligrams { | **Phosphorus** | 25 |
| | **Iron** | .6 |
| | **Sodium** | 18 |
| | **Potassium** | 188 |
| | **Thiamin** | .06 |
| | **Riboflavin** | .04 |
| | **Niacin** | .3 |
| | **Ascorbic Acid** | 43 |
| unit | **Vitamin A** | 30 |

# SAVOY
# CABBAGE

grams    **Weight:**      70

raw, coarsely
shredded or
sliced

| | | |
|---|---|---|
| % | **Water** | 92 |
| | **Calories** | 15 |
| grams | **Protein** | 2 |
| | **Fat** | Trace |
| | **Carbohydrate** | 3 |
| | **Calcium** | 47 |
| milligrams | **Phosphorus** | 38 |
| | **Iron** | .6 |
| | **Sodium** | 15 |
| | **Potassium** | 188 |
| | **Thiamin** | .04 |
| | **Riboflavin** | .06 |
| | **Niacin** | .2 |
| | **Ascorbic Acid** | 39 |
| unit | **Vitamin A** | 140 |

241

# WHITE MUSTARD
# CABBAGE
## (BOKCHOY OR PAKCHOY)

grams **Weight:** 170

cooked,
boiled,
drained, cut
in 1" pieces

| | | |
|---|---|---|
| % | **Water** | 95 |
| | **Calories** | 25 |
| grams | **Protein** | 2 |
| | **Fat** | Trace |
| | **Carbohydrate** | 4 |
| | **Calcium** | 252 |
| | **Phosphorus** | 56 |
| | **Iron** | 1.0 |
| milligrams | **Sodium** | *31 |
| | **Potassium** | 364 |
| | **Thiamin** | .07 |
| | **Riboflavin** | .14 |
| | **Niacin** | 1.2 |
| | **Ascorbic Acid** | 26 |
| unit | **Vitamin A** | 5,270 |

*Value is for unsalted product. If salt is used, increase value by 236 mg. per 100g of vegetable — an estimated figure based on typical amount of salt (0.6%) in canned vegetables.

# CARROTS

| | grams | **Weight:** | 72 | 110 |
|---|---|---|---|---|

| | | raw*<br>whole** | raw*<br>grated |
|---|---|---|---|
| % | **Water** | 88 | 88 |
| | **Calories** | 30 | 45 |
| grams | **Protein** | 1 | 1 |
| | **Fat** | Trace | Trace |
| | **Carbohydrate** | 7 | 11 |
| | **Calcium** | 27 | 40 |
| milligrams | **Phosphorus** | 26 | 40 |
| | **Iron** | .5 | .8 |
| | **Sodium** | 34 | 52 |
| | **Potassium** | 246 | 375 |
| | **Thiamin** | .04 | .07 |
| | **Riboflavin** | .04 | .06 |
| | **Niacin** | .4 | .7 |
| | **Ascorbic Acid** | 6 | 9 |
| unit | **Vitamin A** | 7,930 | 12,100*** |

*Without crowns and tips.
**7½ by 1⅛ in. or strips 2½ to 3 in. long.
***Based on average for carrots marketed as fresh vegetable.

# CARROTS
## COOKED, CANNED

| | grams | Weight: | 155 | 155 | 28 |
|---|---|---|---|---|---|

| | | cooked, (crosswise cuts) drained | canned, sliced, drained solids | canned, strained or junior baby food |
|---|---|---|---|---|
| % | Water | 91 | 91 | 92 |
| grams | Calories | 50 | 45 | 10 |
| | Protein | 1 | 1 | Trace |
| | Fat | Trace | Trace | Trace |
| | Carbohydrate | 11 | 10 | 2 |
| | Calcium | 51 | 47 | 7 |
| milligrams | Phosphorus | 48 | 34 | 6 |
| | Iron | 9 | 1.1 | .1 |
| | Sodium | *51 | *366 | 45 |
| | Potassium | 344 | 186 | 51 |
| | Thiamin | .08 | .03 | .01 |
| | Riboflavin | .08 | .05 | .01 |
| | Niacin | .8 | .6 | .1 |
| | Ascorbic Acid | 9 | 3 | 1 |
| unit | Vitamin A | 16,280 | 23,250 | 3,690 |

*Value is for unsalted product. If salt is used, increase value by 236 mg. per 100g of vegetable — an estimated figure based on typical amount of salt (0.6%) in canned vegetables.

# CAULIFLOWER

| | grams | | |
|---|---|---|---|
| **Weight:** | 115 | 125 | 180 |

| | raw, chopped | cooked, drained: from raw (flower buds) | cooked, drained: from frozen (floweretts) |
|---|---|---|---|
| % **Water** | 91 | 93 | 94 |
| **Calories** | 31 | 30 | 30 |
| **Protein** | 3 | 3 | 3 |
| **Fat** | Trace | Trace | Trace |
| **Carbohydrate** | 6 | 5 | 6 |
| **Calcium** | 29 | 26 | 31 |
| **Phosphorus** | 64 | 53 | 68 |
| **Iron** | 1.3 | .9 | .9 |
| **Sodium** | 15 | *11 | *18 |
| **Potassium** | 339 | 258 | 373 |
| **Thiamin** | .13 | .11 | .07 |
| **Riboflavin** | .12 | .10 | .09 |
| **Niacin** | .8 | .8 | .7 |
| **Ascorbic Acid** | 90 | 69 | .74 |
| **Vitamin A** | 70 | 80 | 50 |

grams / milligrams / unit

*Value is for unsalted product. If salt is used, increase value by 236 mg. per 100g of vegetable — an estimated figure based on typical amount of salt (0.6%) in canned vegetables.

# CELERY
## PASCAL TYPE

| | grams | | |
|---|---|---|---|
| | **Weight:** | 40 | 120 |
| | | raw: stalk, large outer, 8 by 1½″ at root end | raw: pieces diced |
| % | **Water** | 94 | 94 |
| | **Calories** | 5 | 20 |
| grams | **Protein** | Trace | 1 |
| | **Fat** | Trace | Trace |
| | **Carbohydrate** | 2 | 5 |
| | **Calcium** | 16 | 47 |
| milligrams | **Phosphorus** | 11 | 34 |
| | **Iron** | .1 | .4 |
| | **Sodium** | 50 | 151 |
| | **Potassium** | 136 | 409 |
| | **Thiamin** | .01 | .04 |
| | **Riboflavin** | .01 | .04 |
| | **Niacin** | .1 | .4 |
| | **Ascorbic Acid** | 4 | 11 |
| unit | **Vitamin A** | 110 | 320 |

# CHARD, SWISS

| grams | Weight: | 145 | 175 | |
|---|---|---|---|---|
| | | cooked, leaves & stalks | cooked, leaves only | |
| % | **Water** | 93.7 | 93.7 | |
| | **Calories** | 26 | 32 | |
| | **Protein** | 2.6 | 3.2 | |
| grams | **Fat** | .3 | .4 | |
| | **Carbohydrate** | 4.8 | 5.8 | |
| | **Calcium** | 106 | 128 | |
| | **Phosphorus** | 35 | 42 | |
| | **Iron** | 2.6 | 3.2 | |
| | **Sodium** | *125 | *151 | |
| milligrams | **Potassium** | 465 | 562 | |
| | **Thiamin** | .06 | .07 | |
| | **Riboflavin** | .16 | .19 | |
| | **Niacin** | .6 | .7 | |
| | **Ascorbic Acid** | 23 | 28 | |
| unit | **Vitamin A** | 7,830 | 9,450 | |

*Value is for unsalted product. If salt is used, increase value by 236 mg. per 100g of vegetable — an estimated figure based on typical amount of salt (0.6%) in canned vegetables.

# CHICORY, WITLOOF

grams **Weight:** 90

raw,
chopped,
1½" pieces

| | | |
|---|---|---|
| % | Water | 95.1 |
| | Calories | 14 |
| grams | Protein | .9 |
| | Fat | .1 |
| | Carbohydrate | 2.9 |
| | Calcium | 16 |
| | Phosphorus | 19 |
| | Iron | .5 |
| | Sodium | 6 |
| milligrams | Potassium | 164 |
| | Thiamin | N.A. |
| | Riboflavin | N.A. |
| | Niacin | N.A. |
| | Ascorbic Acid | N.A. |
| unit | Vitamin A | Trace |

# CHIVES

grams  **Weight:**          3

raw,
chopped,
⅛" pieces

| % | **Water** | 91.3 |
|---|---|---|
| | **Calories** | 1 |
| | **Protein** | .1 |
| | **Fat** | Trace |
| | **Carbohydrate** | .2 |
| | **Calcium** | 2 |
| | **Phosphorus** | 1 |
| | **Iron** | .1 |
| | **Sodium** | N.A. |
| | **Potassium** | 8 |
| | **Thiamin** | Trace |
| | **Riboflavin** | Trace |
| | **Niacin** | Trace |
| | **Ascorbic Acid** | 2 |
| unit | **Vitamin A** | 170 |

grams — milligrams

249

# COLLARDS

| | grams | Weight: | 190 | 170 |
|---|---|---|---|---|

| | | cooked, drained from raw (leaves without stems) | cooked, drained from frozen (chopped) |
|---|---|---|---|
| % | Water | 90 | 90 |
| | Calories | 65 | 50 |
| grams | Protein | 7 | 5 |
| | Fat | 1 | 1 |
| | Carbohydrate | 10 | 10 |
| | Calcium | 357 | 299 |
| milligrams | Phosphorus | 99 | 87 |
| | Iron | 1.5 | 1.7 |
| | Sodium | *20 | *27 |
| | Potassium | 498 | 401 |
| | Thiamin | .21 | .10 |
| | Riboflavin | .38 | .24 |
| | Niacin | 2.3 | 1.0 |
| | Ascorbic Acid | 144 | 56 |
| unit | Vitamin A | 14,820 | 11,560 |

*Value is for unsalted product. If salt is used, increase value by 236 mg. per 100g of vegetable — an estimated figure based on typical amount of salt (0.6%) in canned vegetables.

250

# CORN
## COOKED

| | grams Weight: | 140 | 229 | 165 |
|---|---|---|---|---|
| | | cooked from raw ear, 5 by 1¼"* | cooked from frozen ear, 5" long* | cooked from frozen: kernels |
| % | Water | 74 | 73 | 77 |
| | Calories | 70 | 120 | 130 |
| grams | Protein | 2 | 4 | 5 |
| | Fat | 1 | 1 | 1 |
| | Carbohydrate | 16 | 27 | 31 |
| | Calcium | 2 | 4 | 5 |
| milligrams | Phosphorus | 69 | 121 | 120 |
| | Iron | .5 | 1.0 | 1.3 |
| | Sodium | ***Trace | ***1 | ***2 |
| | Potassium | 151 | 291 | 304 |
| | Thiamin | .09 | .18 | .15 |
| | Riboflavin | .08 | .10 | .10 |
| | Niacin | 1.1 | 2.1 | 2.5 |
| | Ascorbic Acid | 7 | 9 | 8 |
| unit | Vitamin A | **310 | **440 | **580 |

*Weight includes cob. Without cob weight is 77g for raw, 126g for frozen.
**Based on yellow varieties. For white varieties, value is trace.
***Value is for unsalted product. If salt is used, increase value by 236 mg. per 100g of vegetable—an estimated figure based on typical amount of salt (0.6%) in canned vegetables.

# CORN

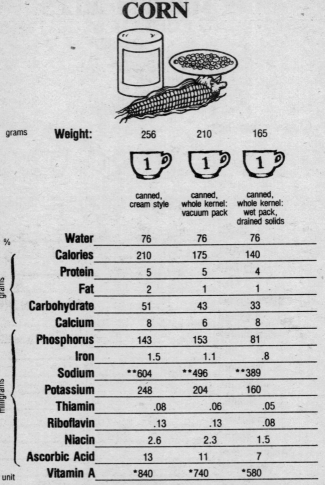

| | | canned, cream style | canned, whole kernel: vacuum pack | canned, whole kernel: wet pack, drained solids |
|---|---|---|---|---|
| grams | **Weight:** | 256 | 210 | 165 |
| % | **Water** | 76 | 76 | 76 |
| | **Calories** | 210 | 175 | 140 |
| grams | **Protein** | 5 | 5 | 4 |
| | **Fat** | 2 | 1 | 1 |
| | **Carbohydrate** | 51 | 43 | 33 |
| | **Calcium** | 8 | 6 | 8 |
| milligrams | **Phosphorus** | 143 | 153 | 81 |
| | **Iron** | 1.5 | 1.1 | .8 |
| | **Sodium** | **604 | **496 | **389 |
| | **Potassium** | 248 | 204 | 160 |
| | **Thiamin** | .08 | .06 | .05 |
| | **Riboflavin** | .13 | .13 | .08 |
| | **Niacin** | 2.6 | 2.3 | 1.5 |
| | **Ascorbic Acid** | 13 | 11 | 7 |
| unit | **Vitamin A** | *840 | *740 | *580 |

*Based on yellow varieties. For white varieties, value is trace.
**Estimated value based on addition of salt in amount of 0.6% of finished product.

252

# CUCUMBER PICKLES

| grams | **Weight:** | 65 | 15 |
|---|---|---|---|
| | |  |  |
| | | dill, medium, whole* | sweet, gherkin, small, whole** |

| | | | |
|---|---|---|---|
| % | **Water** | 93 | 61 |
| | **Calories** | 5 | 20 |
| grams { | **Protein** | Trace | Trace |
| | **Fat** | Trace | Trace |
| | **Carbohydrate** | 1 | 5 |
| | **Calcium** | 17 | 2 |
| milligrams { | **Phosphorus** | 14 | 2 |
| | **Iron** | .7 | .2 |
| | **Sodium** | 928 | 101 |
| | **Potassium** | 130 | N.A. |
| | **Thiamin** | Trace | Trace |
| | **Riboflavin** | .01 | Trace |
| | **Niacin** | Trace | Trace |
| | **Ascorbic Acid** | 4 | 1 |
| unit | **Vitamin A** | 170 | 10 |

*3¾″ long, 1¼″ diam.
**2½″ long, ¾″ diam.

# CUCUMBER PICKLES
## (PACKED, RELISH)

| | grams | Weight: | 15 | 15 |
|---|---|---|---|---|
| | | | fresh pack, slices 1½" diam. ¼" | relish, finely chopped, sweet |
| % | Water | | 79 | 63 |
| grams | Calories | | 10 | 20 |
| | Protein | | Trace | Trace |
| | Fat | | Trace | Trace |
| | Carbohydrate | | 3 | 5 |
| | Calcium | | 5 | 3 |
| milligrams | Phosphorus | | 4 | 2 |
| | Iron | | .3 | .1 |
| | Sodium | | 101 | 107 |
| | Potassium | | N.A. | N.A. |
| | Thiamin | | Trace | N.A. |
| | Riboflavin | | Trace | Trace |
| | Niacin | | Trace | N.A. |
| | Ascorbic Acid | | 1 | N.A. |
| unit | Vitamin A | | 20 | N.A. |

# CUCUMBER SLICES

| | grams | **Weight:** | 28 | 28 |
|---|---|---|---|---|

| | | raw, with peel* | raw without peel* |
|---|---|---|---|
| % | **Water** | 95 | 96 |
| | **Calories** | 5 | 5 |
| grams | **Protein** | Trace | Trace |
| | **Fat** | Trace | Trace |
| | **Carbohydrate** | 1 | 1 |
| | **Calcium** | 7 | 5 |
| milligrams | **Phosphorus** | 8 | 5 |
| | **Iron** | .3 | .1 |
| | **Sodium** | 2 | 2 |
| | **Potassium** | 45 | 45 |
| | **Thiamin** | .01 | .01 |
| | **Riboflavin** | .01 | .01 |
| | **Niacin** | .1 | .1 |
| | **Ascorbic Acid** | 3 | 3 |
| unit | **Vitamin A** | 70 | Trace |

*⅛″ thick (large, 2⅛″ diam:, small 1¾″ diam.) (with peel — 6 large/8 small)
(without peel — 6½ large/9 small).

# DANDELION GREENS

grams  **Weight:**  105

cooked,
drained

| | | |
|---|---|---|
| % | **Water** | 90 |
| | **Calories** | 35 |
| grams | **Protein** | 2 |
| | **Fat** | 1 |
| | **Carbohydrate** | 7 |
| | **Calcium** | 147 |
| | **Phosphorus** | 44 |
| | **Iron** | 1.9 |
| milligrams | **Sodium** | *46 |
| | **Potassium** | 244 |
| | **Thiamin** | .14 |
| | **Riboflavin** | .17 |
| | **Niacin** | N.A. |
| | **Ascorbic Acid** | 19 |
| unit | **Vitamin A** | 12,290 |

*Value is for unsalted product. If salt is used, increase value by 236 mg. per
100g of vegetable — an estimated figure based on typical amount of salt (0.6%)
in canned vegetables.

# ENDIVE, CURLY
## (INCLUDING ESCAROLE)

grams **Weight:** 50

raw, small
pieces

| | | |
|---|---|---|
| % | Water | 93 |
| grams | Calories | 10 |
| | Protein | 1 |
| | Fat | Trace |
| | Carbohydrate | 2 |
| | Calcium | 41 |
| milligrams | Phosphorus | 27 |
| | Iron | .9 |
| | Sodium | 7 |
| | Potassium | 147 |
| | Thiamin | .04 |
| | Riboflavin | .07 |
| | Niacin | .3 |
| | Ascorbic Acid | 5 |
| unit | Vitamin A | 1,650 |

257

# KALE

| | | | |
|---|---|---|---|
| grams | **Weight:** | 110 | 130 |

| | | cooked, drained, from raw (leaves without stems & mid-ribs) | cooked, drained, from frozen (leaf style) |
|---|---|---|---|
| % | **Water** | 88 | 91 |
| | **Calories** | 45 | 40 |
| grams | **Protein** | 5 | 4 |
| | **Fat** | 1 | 1 |
| | **Carbohydrate** | 7 | 7 |
| | **Calcium** | 206 | 157 |
| milligrams | **Phosphorus** | 64 | 62 |
| | **Iron** | 1.8 | 1.3 |
| | **Sodium** | *47 | *27 |
| | **Potassium** | 243 | 251 |
| | **Thiamin** | .11 | .08 |
| | **Riboflavin** | .20 | .20 |
| | **Niacin** | 1.8 | .9 |
| | **Ascorbic Acid** | 102 | 49 |
| unit | **Vitamin A** | 9,130 | 10,660 |

*Value is for unsalted product. If salt is used, increase value by 236 mg. per 100g of vegetable — an estimated figure based on typical amount of salt (0.6%) in canned vegetables.

# BUTTERHEAD VARIETIES
## LETTUCE
### (SUCH AS BOSTON & OAKLEAF)

grams     **Weight:**     *220    **15

( 1 )

|  | Butterhead, head 5" diam. | Butterhead leaves |
|---|---|---|
| % | **Water** | 95 | 95 |
| **Calories** | 25 | Trace |
| **Protein** | 2 | Trace |
| **Fat** | Trace | Trace |
| **Carbohydrate** | 4 | Trace |
| **Calcium** | 57 | 5 |
| **Phosphorus** | 42 | 4 |
| **Iron** | 3.3 | .3 |
| **Sodium** | 15 | 1 |
| **Potassium** | 430 | 40 |
| **Thiamin** | .10 | .01 |
| **Riboflavin** | .10 | .01 |
| **Niacin** | .5 | Trace |
| **Ascorbic Acid** | 13 | 1 |
| **Vitamin A** | 1,580 | 150 |

*grams* — Water through Carbohydrate (%), Calcium through Vitamin A (milligrams), Vitamin A (unit)

*A head includes refuse of outer leaves and core. Without these parts, weight is 163g.
**1 outer, or 2 inner, or 3 heart leaves.

# CRISPHEAD VARIETIES
# LETTUCE
## (SUCH AS ICEBERG)

| grams | Weight: | *567 | 135 | 55 |
|---|---|---|---|---|
| | |  | |  |
| | | head 6" diam. | wedge ¼ of head | pieces, chopped or shredded |
| % | **Water** | 96 | 96 | 96 |
| | **Calories** | 70 | 20 | 5 |
| grams | **Protein** | 5 | 1 | Trace |
| | **Fat** | 1 | Trace | Trace |
| | **Carbohydrate** | 16 | 4 | 2 |
| | **Calcium** | 108 | 27 | 11 |
| milligrams | **Phosphorus** | 118 | 30 | 12 |
| | **Iron** | 2.7 | .7 | .3 |
| | **Sodium** | 48 | 12 | 5 |
| | **Potassium** | 943 | 236 | 96 |
| | **Thiamin** | .32 | .08 | .03 |
| | **Riboflavin** | .32 | .08 | .03 |
| | **Niacin** | 1.6 | .4 | .2 |
| | **Ascorbic Acid** | 32 | 8 | 3 |
| unit | **Vitamin A** | 1,780 | 450 | 180 |

*Weight includes core. Without core, weight is 539g.

# LOOSELEAF OR BUNCHING
# LETTUCE
## (SUCH AS ROMAINE & COS)

grams    **Weight:**    55

looseleaf
chopped or
shredded
pieces

| | | |
|---|---|---|
| % | **Water** | 94 |
| | **Calories** | 10 |
| | **Protein** | 1 |
| | **Fat** | Trace |
| | **Carbohydrate** | 2 |
| | **Calcium** | 37 |
| | **Phosphorus** | 14 |
| | **Iron** | .8 |
| | **Sodium** | 5 |
| | **Potassium** | 145 |
| | **Thiamin** | .03 |
| | **Riboflavin** | .04 |
| | **Niacin** | .2 |
| | **Ascorbic Acid** | 10 |
| unit | **Vitamin A** | 1,050 |

grams { (Protein, Fat, Carbohydrate)

milligrams { (Calcium through Ascorbic Acid)

261

# MIXED VEGETABLES

grams  **Weight:**        182

frozen,
cooked

| | | |
|---|---|---|
| %  | **Water** | 83 |
| | **Calories** | 116 |
| | **Protein** | 6 |
| | **Fat** | 1 |
| | **Carbohydrate** | 24 |
| | **Calcium** | 46 |
| | **Phosphorus** | 115 |
| | **Iron** | 2.4 |
| | **Sodium** | *96 |
| | **Potassium** | 348 |
| | **Thiamin** | .22 |
| | **Riboflavin** | .13 |
| | **Niacin** | 2.0 |
| | **Ascorbic Acid** | 15 |
| unit | **Vitamin A** | 9,010 |

grams { Calories … Calcium }
milligrams { Phosphorus … Ascorbic Acid }

*Value based on average weighted in accordance with commercial practice in freezing vegetables. For cooked vegetables, value also represents no additional salting. If salt is moderately added, increase value by 236 mg. per 100g. − an estimate based on typical amount of salt (0.6%) in canned vegetables.

262

# MUSHROOMS

grams    **Weight:**        70

sliced or
chopped

| | | |
|---|---|---|
| % | **Water** | 90 |
| | **Calories** | 20 |
| | **Protein** | 2 |
| | **Fat** | Trace |
| | **Carbohydrate** | 3 |
| | **Calcium** | 4 |
| | **Phosphorus** | 81 |
| | **Iron** | .6 |
| | **Sodium** | 11 |
| | **Potassium** | 290 |
| | **Thiamin** | .07 |
| | **Riboflavin** | .32 |
| | **Niacin** | 2.9 |
| | **Ascorbic Acid** | 2 |
| unit | **Vitamin A** | Trace |

(grams, milligrams)

# MUSTARD GREENS

| | | |
|---|---|---|
| grams | **Weight:** | 140 |

cooked,
drained,
without stems
& midribs

| | | |
|---|---|---|
| % | **Water** | 93 |
| | **Calories** | 30 |
| grams { | **Protein** | 3 |
| | **Fat** | 1 |
| | **Carbohydrate** | 6 |
| | **Calcium** | 193 |
| milligrams { | **Phosphorus** | 45 |
| | **Iron** | 2.5 |
| | **Sodium** | *25 |
| | **Potassium** | 308 |
| | **Thiamin** | .11 |
| | **Riboflavin** | .20 |
| | **Niacin** | .8 |
| | **Ascorbic Acid** | 67 |
| unit | **Vitamin A** | 8,120 |

*Value is for unsalted product. If salt is used, increase value by 236 mg. per
100g of vegetable — an estimated figure based on typical amount of salt (0.6%)
in canned vegetables.

# OKRA PODS

grams   **Weight:**        106

**10**

3 by ⅝″
cooked

| | % |
|---|---|
| **Water** | 91 |
| **Calories** | 30 |
| **Protein** | 2 |
| **Fat** | Trace |
| **Carbohydrate** | 6 |
| **Calcium** | 98 |
| **Phosphorus** | 43 |
| **Iron** | .5 |
| **Sodium** | *2 |
| **Potassium** | 184 |
| **Thiamin** | .14 |
| **Riboflavin** | .19 |
| **Niacin** | 1.0 |
| **Ascorbic Acid** | 21 |
| **Vitamin A** | 520 |

grams (Water through Carbohydrate)
milligrams (Calcium through Ascorbic Acid)
unit (Vitamin A)

*Value is for unsalted product. If salt is used, increase value by 236 mg. per
100g of vegetable — an estimated figure based on typical amount of salt (0.6%)
in canned vegetables.

# OLIVES

| grams | Weight: | *34 | *34 |
|---|---|---|---|

| | | pickled green | pickled ripe Mission |
|---|---|---|---|
| % | Water | 78.2 | 73.0 |
| grams | Calories | 33 | 54 |
| | Protein | .4 | .4 |
| | Fat | 3.6 | 5.9 |
| | Carbohydrate | .4 | .9 |
| | Calcium | 17 | 31 |
| milligrams | Phosphorus | 5 | 5 |
| | Iron | .5 | .5 |
| | Sodium | 686 | 219 |
| | Potassium | 16 | 8 |
| | Thiamin | N.A. | Trace |
| | Riboflavin | N.A. | N.A. |
| | Niacin | N.A. | N.A. |
| | Ascorbic Acid | N.A. | N.A. |
| unit | Vitamin A | 90 | 20 |

*10 small (approximately 10/16 in. diam — 13/16 in. long).

# ONIONS, GREEN

| | grams | Weight: | 30 |

| | | Bulb* and white portion of top |
|---|---|---|
| % | **Water** | 88 |
| | **Calories** | 15 |
| **grams** { | **Protein** | Trace |
| | **Fat** | Trace |
| | **Carbohydrate** | 3 |
| | **Calcium** | 12 |
| **milligrams** { | **Phosphorus** | 12 |
| | **Iron** | .2 |
| | **Sodium** | 2 |
| | **Potassium** | 69 |
| | **Thiamin** | .02 |
| | **Riboflavin** | .01 |
| | **Niacin** | .1 |
| | **Ascorbic Acid** | 8 |
| unit | **Vitamin A** | Trace |

*³⁄₈" diam.

# ONIONS, MATURE

| | raw: chopped | raw: sliced | cooked (whole or sliced) drained |
|---|---|---|---|
| grams **Weight:** | 170 | 115 | 210 |

| | | | | |
|---|---|---|---|---|
| % | **Water** | 89 | 89 | 92 |
| grams | **Calories** | 65 | 45 | 60 |
| | **Protein** | 3 | 2 | 3 |
| | **Fat** | Trace | Trace | Trace |
| | **Carbohydrate** | 15 | 10 | 14 |
| milligrams | **Calcium** | 46 | 31 | 50 |
| | **Phosphorus** | 61 | 41 | 61 |
| | **Iron** | .9 | .6 | .8 |
| | **Sodium** | 17 | 12 | **15 |
| | **Potassium** | 267 | 181 | 231 |
| | **Thiamin** | .05 | .03 | .06 |
| | **Riboflavin** | .07 | .05 | .06 |
| | **Niacin** | .3 | .2 | .4 |
| | **Ascorbic Acid** | 17 | 12 | 15 |
| unit | **Vitamin A** | *Trace | *Trace | *Trace |

*Value based on white-fleshed varieties. For yellow-fleshed varieties, value in International Units (I.U.) is 70g for chopped, 50 for sliced, and 80 for cooked.
**Value is for unsalted product.

# PARSLEY

grams     **Weight:**     4

raw
chopped

| | | |
|---|---|---|
| % | **Water** | 85 |
| | **Calories** | Trace |
| | **Protein** | Trace |
| grams | **Fat** | Trace |
| | **Carbohydrate** | Trace |
| | **Calcium** | 7 |
| | **Phosphorus** | 2 |
| | **Iron** | .2 |
| | **Sodium** | 2 |
| milligrams | **Potassium** | 25 |
| | **Thiamin** | Trace |
| | **Riboflavin** | .01 |
| | **Niacin** | Trace |
| | **Ascorbic Acid** | 6 |
| unit | **Vitamin A** | 300 |

# PARSNIPS

grams **Weight:** 155

cooked (diced
or 2"
lengths)

| | | |
|---|---|---|
| % | **Water** | 82 |
| | **Calories** | 100 |
| grams | **Protein** | 2 |
| | **Fat** | 1 |
| | **Carbohydrate** | 23 |
| | **Calcium** | 70 |
| milligrams | **Phosphorus** | 96 |
| | **Iron** | .9 |
| | **Sodium** | *12 |
| | **Potassium** | 587 |
| | **Thiamin** | .11 |
| | **Riboflavin** | .12 |
| | **Niacin** | .2 |
| | **Ascorbic Acid** | 16 |
| unit | **Vitamin A** | 50 |

*Value is for unsalted product. If salt is used, increase value by 236 mg. per
100g of vegetable — an estimated figure based on typical amount of salt (0.6%)
in canned vegetables.

# PEAS, GREEN

| | grams | Weight: | 170 | *28 | 160 |
|---|---|---|---|---|---|

| | | canned: whole, drained solids | canned: strained (baby food) | frozen cooked, drained |
|---|---|---|---|---|
| % | Water | 77 | 86 | 82 |
| | Calories | 150 | 15 | 110 |
| grams | Protein | 8 | 1 | 8 |
| | Fat | 1 | Trace | Trace |
| | Carbohydrate | 29 | 3 | 19 |
| | Calcium | 44 | 3 | 30 |
| milligrams | Phosphorus | 129 | 18 | 138 |
| | Iron | 3.2 | .3 | 3.0 |
| | Sodium | **401 | N.A. | **184 |
| | Potassium | 163 | 28 | 216 |
| | Thiamin | .15 | .02 | .43 |
| | Riboflavin | .10 | .03 | .14 |
| | Niacin | 1.4 | .3 | 2.7 |
| | Ascorbic Acid | 14 | 3 | 21 |
| unit | Vitamin A | 1,170 | 140 | 960 |

*1¾ — 2 tbs.
**Estimated value based on addition of salt in amount of 0.6% of finished product.

# PEPPERS, HOT

| grams | **Weight:** | 2 |
|---|---|---|

dried*

| | | |
|---|---|---|
| % | **Water** | 9 |
| | **Calories** | 5 |
| grams { | **Protein** | Trace |
| | **Fat** | Trace |
| | **Carbohydrate** | 1 |
| | **Calcium** | 5 |
| milligrams { | **Phosphorus** | 4 |
| | **Iron** | .3 |
| | **Sodium** | 31 |
| | **Potassium** | 20 |
| | **Thiamin** | Trace |
| | **Riboflavin** | .02 |
| | **Niacin** | .2 |
| | **Ascorbic Acid** | Trace |
| unit | **Vitamin A** | 1,300 |

*Without seeds, ground chili powder, added seasonings.

# PEPPERS, SWEET

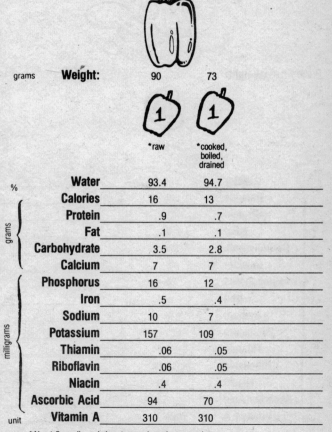

| | | *raw | *cooked, boiled, drained |
|---|---|---|---|
| grams | **Weight:** | 90 | 73 |
| % | **Water** | 93.4 | 94.7 |
| | **Calories** | 16 | 13 |
| grams | **Protein** | .9 | .7 |
| | **Fat** | .1 | .1 |
| | **Carbohydrate** | 3.5 | 2.8 |
| | **Calcium** | 7 | 7 |
| milligrams | **Phosphorus** | 16 | 12 |
| | **Iron** | .5 | .4 |
| | **Sodium** | 10 | 7 |
| | **Potassium** | 157 | 109 |
| | **Thiamin** | .06 | .05 |
| | **Riboflavin** | .06 | .05 |
| | **Niacin** | .4 | .4 |
| | **Ascorbic Acid** | 94 | 70 |
| unit | **Vitamin A** | 310 | 310 |

*About 5 per lb., whole, stem and seeds removed.

# POTATO CHIPS

grams  **Weight:**  20

**10**

*10 chips

| | | |
|---|---|---|
| % | **Water** | 2 |
| grams | **Calories** | 115 |
| | **Protein** | 1 |
| | **Fat** | 8 |
| | **Carbohydrate** | 10 |
| | **Calcium** | 8 |
| milligrams | **Phosphorus** | 28 |
| | **Iron** | .4 |
| | **Sodium** | **N.A. |
| | **Potassium** | 226 |
| | **Thiamin** | .04 |
| | **Riboflavin** | .01 |
| | **Niacin** | 1.0 |
| | **Ascorbic Acid** | .3 |
| unit | **Vitamin A** | Trace |

*1¾ by 2½ in.
**Sodium content is variable and may be as high as 1,000 mg. per 100g.

# POTATO SALAD

| | | |
|---|---|---|
| grams | **Weight:** | 250 |

made with
cooked salad
dressing

| | | |
|---|---|---|
| % | **Water** | 76 |
| | **Calories** | 250 |
| grams | **Protein** | 7 |
| | **Fat** | 7 |
| | **Carbohydrate** | 41 |
| | **Calcium** | 80 |
| milligrams | **Phosphorus** | 160 |
| | **Iron** | 1.5 |
| | **Sodium** | 1,320 |
| | **Potassium** | 798 |
| | **Thiamin** | .20 |
| | **Riboflavin** | .18 |
| | **Niacin** | 2.8 |
| | **Ascorbic Acid** | 28 |
| unit | **Vitamin A** | 350 |

# POTATOES*
## BAKED

grams **Weight:** 202

peeled after
baking

| | % | |
|---|---|---|
| Water | 75.1 | |
| Calories | 145 | |
| Protein | 4.0 | |
| Fat | .2 | |
| Carbohydrate | 32.8 | |
| Calcium | 14 | |
| Phosphorus | 101 | |
| Iron | 1.1 | |
| Sodium | **6 | |
| Potassium | 782 | |
| Thiamin | .15 | |
| Riboflavin | .07 | |
| Niacin | 2.7 | |
| Ascorbic Acid | ***.31 | |
| Vitamin A | Trace | |

*about 2 potatoes per lb. raw.
**Value is for unsalted product.
***Based on year round average, recently dug potatoes have more than stored potatoes.

# POTATOES
## BOILED

| | | peeled after boiling* | peeled before boiling* |
|---|---|---|---|
| grams | Weight: | 250 | 188 |
| | |  |  |
| % | Water | 79.8 | 82.8 |
| grams | Calories | 173 | 122 |
| | Protein | 4.8 | 3.6 |
| | Fat | .2 | .2 |
| | Carbohydrate | 38.9 | 27.3 |
| | Calcium | 16 | 11 |
| milligrams | Phosphorus | 121 | 79 |
| | Iron | 1.4 | .9 |
| | Sodium | **7 | **4 |
| | Potassium | 926 | 536 |
| | Thiamin | .20 | .17 |
| | Riboflavin | .09 | .07 |
| | Niacin | 3.4 | 2.3 |
| | Ascorbic Acid | ***36 | ***30 |
| unit | Vitamin A | Trace | Trace |

*about 3 potatoes per lb. raw.
**Value is for unsalted product.
***Based on year round average, recently dug potatoes have more than stored potatoes.

# POTATOES
## FRENCH FRIED

| | grams Weight: | 50 | 50 |
|---|---|---|---|
| | | strips 2"-3½" prepared from raw | strips 2"-3½" frozen oven heated |
| % | Water | 45 | 53 |
| | Calories | 135 | 110 |
| grams | Protein | 2 | 2 |
| | Fat | 7 | 4 |
| | Carbohydrate | 18 | 17 |
| | Calcium | 8 | 5 |
| milligrams | Phosphorus | 56 | 43 |
| | Iron | .7 | .9 |
| | Sodium | *3 | *2 |
| | Potassium | 427 | 326 |
| | Thiamin | .07 | .07 |
| | Riboflavin | .04 | .01 |
| | Niacin | 1.6 | 1.3 |
| | Ascorbic Acid | 11 | 11 |
| unit | Vitamin A | Trace | Trace |

*Value is for unsalted product. If salt is used, increase value by 236 mg. per 100g of vegetable.

278

# POTATOES
## HASHED BROWN

grams **Weight:** 155

prepared
from frozen

| | | |
|---|---|---|
| % | **Water** | 56 |
| | **Calories** | 345 |
| grams { | **Protein** | 3 |
| | **Fat** | 18 |
| | **Carbohydrate** | 45 |
| | **Calcium** | 28 |
| milligrams { | **Phosphorus** | 78 |
| | **Iron** | 1.9 |
| | **Sodium** | 463 |
| | **Potassium** | 439 |
| | **Thiamin** | .11 |
| | **Riboflavin** | .03 |
| | **Niacin** | 1.6 |
| | **Ascorbic Acid** | 12 |
| unit | **Vitamin A** | Trace |

# POTATOES
## MASHED

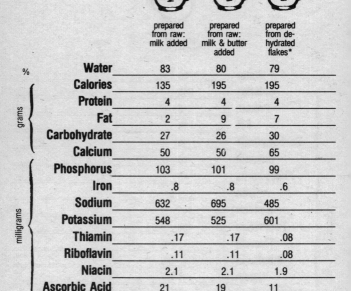

| | prepared from raw: milk added | prepared from raw: milk & butter added | prepared from de-hydrated flakes* |
|---|---|---|---|
| **Weight:** grams | 210 | 210 | 210 |
| **Water** % | 83 | 80 | 79 |
| **Calories** | 135 | 195 | 195 |
| **Protein** | 4 | 4 | 4 |
| **Fat** | 2 | 9 | 7 |
| **Carbohydrate** | 27 | 26 | 30 |
| **Calcium** | 50 | 50 | 65 |
| **Phosphorus** | 103 | 101 | 99 |
| **Iron** | .8 | .8 | .6 |
| **Sodium** | 632 | 695 | 485 |
| **Potassium** | 548 | 525 | 601 |
| **Thiamin** | .17 | .17 | .08 |
| **Riboflavin** | .11 | .11 | .08 |
| **Niacin** | 2.1 | 2.1 | 1.9 |
| **Ascorbic Acid** | 21 | 19 | 11 |
| **Vitamin A** unit | 40 | 360 | 270 |

*(without milk) water, milk, butter, and salt added

# PUMPKIN

| | | |
|---|---|---|
| grams | **Weight:** | 245 |

canned

| % | **Water** | 90 | |
|---|---|---|---|
| | **Calories** | 80 | |
| | **Protein** | 2 | |
| grams | **Fat** | 1 | |
| | **Carbohydrate** | 19 | |
| | **Calcium** | 61 | |
| | **Phosphorus** | 64 | |
| | **Iron** | 1.0 | |
| | **Sodium** | *5 | |
| milligrams | **Potassium** | 588 | |
| | **Thiamin** | .07 | |
| | **Riboflavin** | .12 | |
| | **Niacin** | 1.5 | |
| | **Ascorbic Acid** | 12 | |
| unit | **Vitamin A** | 15,680 | |

*Applies to product without salt added.

# RADISHES

grams     **Weight:**     *90

stem ends,
rootlets cut
off

| | | |
|---|---|---|
| % | **Water** | 94.5 |
| | **Calories** | 14 |
| grams | **Protein** | .8 |
| | **Fat** | .1 |
| | **Carbohydrate** | 2.9 |
| | **Calcium** | 24 |
| | **Phosphorus** | 25 |
| | **Iron** | .8 |
| | **Sodium** | 15 |
| milligrams | **Potassium** | 261 |
| | **Thiamin** | .02 |
| | **Riboflavin** | .02 |
| | **Niacin** | .2 |
| | **Ascorbic Acid** | 21 |
| unit | **Vitamin A** | 10 |

*10 radishes.

282

# SAUERKRAUT

grams **Weight:** 235

canned,
solids and
liquid

| | | |
|---|---|---|
| **Water** | 93 | % |
| **Calories** | 40 | |
| **Protein** | 2 | |
| **Fat** | Trace | |
| **Carbohydrate** | 9 | |
| **Calcium** | 85 | |
| **Phosphorus** | 41 | |
| **Iron** | 1.2 | |
| **Sodium** | *1,755 | |
| **Potassium** | 329 | |
| **Thiamin** | .07 | |
| **Riboflavin** | .09 | |
| **Niacin** | .5 | |
| **Ascorbic Acid** | 33 | |
| **Vitamin A** | 120 | |

grams { Water, Calories, Protein, Fat, Carbohydrate, Calcium }
milligrams { Phosphorus ... Ascorbic Acid }
unit { Vitamin A }

*Values for sauerkraut and sauerkraut juice are based on salt contents of 1.9 and 2.0% respectively, in finished products. Amounts in some samples may vary significantly from this estimate.

283

# SUMMER VARIETIES
# SQUASH
## (ZUCCHINI, CROOKNECK, ETC.)

grams **Weight:** 210

1

diced
cooked,
drained

| | |
|---|---|
| % | |
| **Water** | 95.5 |
| **Calories** | 29 |
| **Protein** | 1.9 |
| **Fat** | .2 |
| **Carbohydrate** | 6.5 |
| **Calcium** | 53 |
| **Phosphorus** | 53 |
| **Iron** | .8 |
| **Sodium** | *2 |
| **Potassium** | 296 |
| **Thiamin** | .11 |
| **Riboflavin** | .17 |
| **Niacin** | 1.7 |
| **Ascorbic Acid** | 21 |
| **Vitamin A** | 820 |

grams (Protein through Calcium), milligrams (Calcium through Vitamin A), unit (Vitamin A)

*Value is for unsalted product. If salt is used, increase value by 236 mg. per 100g of vegetable — an estimated figure based on typical amount of salt (0.6%) in canned vegetables.

# WINTER VARIETIES
# SQUASH
## (ACORN, BUTTERNUT, ETC.)

grams     **Weight:**      205

baked,
mashed

| | | |
|---|---|---|
| % | **Water** | 81 |
| | **Calories** | 130 |
| grams | **Protein** | 4 |
| | **Fat** | 1 |
| | **Carbohydrate** | 32 |
| | **Calcium** | 57 |
| milligrams | **Phosphorus** | 98 |
| | **Iron** | 1.6 |
| | **Sodium** | *2 |
| | **Potassium** | 945 |
| | **Thiamin** | .10 |
| | **Riboflavin** | .27 |
| | **Niacin** | 1.4 |
| | **Ascorbic Acid** | 27 |
| unit | **Vitamin A** | 8,610 |

*Value is for unsalted product. If salt is used, increase value by 236 mg. per 100g of vegetable.

# SPINACH

| | grams | | |
|---|---|---|---|
| | **Weight:** | 55 | 180 |

| | | raw, chopped | cooked, drained: from raw |
|---|---|---|---|
| % | **Water** | 91 | 92 |
| grams | **Calories** | 15 | 40 |
| | **Protein** | 2 | 5 |
| | **Fat** | Trace | 1 |
| | **Carbohydrate** | 2 | 6 |
| milligrams | **Calcium** | 51 | 167 |
| | **Phosphorus** | 28 | 68 |
| | **Iron** | 1.7 | 4.0 |
| | **Sodium** | 39 | *90 |
| | **Potassium** | 259 | 583 |
| | **Thiamin** | .06 | .13 |
| | **Riboflavin** | .11 | .25 |
| | **Niacin** | .3 | .9 |
| | **Ascorbic Acid** | 28 | 50 |
| unit | **Vitamin A** | 4,460 | 14,580 |

*Value is for unsalted product. If salt is used, increase value by 236 mg. per 100g of vegetable.

# SPINACH
## (FROZEN, CANNED)

| | | cooked, drained: from frozen: chopped | canned drained solids |
|---|---|---|---|
| grams | **Weight:** | 205 | 205 |
| % | **Water** | 92 | 91 |
| | **Calories** | 45 | 50 |
| grams | **Protein** | 6 | 6 |
| | **Fat** | 1 | 1 |
| | **Carbohydrate** | 8 | 7 |
| | **Calcium** | 232 | 242 |
| milligrams | **Phosphorus** | 90 | 5.3 |
| | **Iron** | 4.3 | 5.3 |
| | **Sodium** | *707 | 66 |
| | **Potassium** | 683 | 513 |
| | **Thiamin** | .14 | .04 |
| | **Riboflavin** | .31 | .25 |
| | **Niacin** | .8 | .6 |
| | **Ascorbic Acid** | 39 | 29 |
| unit | **Vitamin A** | 16,200 | 16,400 |

*Value is for unsalted product. If salt is used, increase value by 236 mg. per 100g of vegetable.

# SWEET POTATOES

| | grams | **Weight:** | 146 | 180 |

| | | *baked in skin, peeled | boiled in skin, peeled |
|---|---|---|---|
| % | **Water** | 64 | 71 |
| | **Calories** | 161 | 172 |
| grams | **Protein** | 2 | 2.6 |
| | **Fat** | .6 | .6 |
| | **Carbohydrate** | 37 | 39.8 |
| | **Calcium** | 46 | 48 |
| milligrams | **Phosphorus** | 66 | 71 |
| | **Iron** | 1.0 | 1.1 |
| | **Sodium** | **14 | **15 |
| | **Potassium** | 342 | 367 |
| | **Thiamin** | .10 | .14 |
| | **Riboflavin** | .08 | .09 |
| | **Niacin** | .8 | .9 |
| | **Ascorbic Acid** | 25 | 26 |
| unit | **Vitamin A** | 9,230 | 11,940 |

*Raw, 5 by 2″, about 2⅓ per lb.
**Value is for unsalted product.

# SWEET POTATOES
## (CANNED)

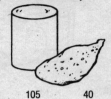

| | grams | Weight: | 105 | 40 |
|---|---|---|---|---|

candied*  canned**
vacuum pack

| | | candied* | canned** vacuum pack |
|---|---|---|---|
| % | Water | 60 | 72 |
| | Calories | 175 | 45 |
| grams | Protein | 1 | 1 |
| | Fat | 3 | Trace |
| | Carbohydrate | 36 | 10 |
| | Calcium | 39 | 10 |
| milligrams | Phosphorus | 45 | 16 |
| | Iron | .9 | .3 |
| | Sodium | 44 | 19 |
| | Potassium | 200 | 80 |
| | Thiamin | .06 | .02 |
| | Riboflavin | .04 | .02 |
| | Niacin | .4 | .2 |
| | Ascorbic Acid | 11 | 6 |
| unit | Vitamin A | 6,620 | 3,120 |

*2½ by 2" piece.
**Value is for unsalted product.

# TOMATOES

| | | grams | **Weight:** | 135 | 241 |

| | | raw<br>2⅗″ diam.<br>cores and stem | canned,<br>solids and<br>liquid |
|---|---|---|---|
| % | **Water** | 94 | 94 |
| | **Calories** | 25 | 50 |
| grams | **Protein** | 1 | 2 |
| | **Fat** | Trace | Trace |
| | **Carbohydrate** | 6 | 10 |
| | **Calcium** | 16 | **14 |
| milligrams | **Phosphorus** | 33 | 46 |
| | **Iron** | .6 | 1.2 |
| | **Sodium** | 4 | 313 |
| | **Potassium** | 300 | 523 |
| | **Thiamin** | .07 | .12 |
| | **Riboflavin** | .05 | .07 |
| | **Niacin** | .9 | 1.7 |
| | **Ascorbic Acid** | *28 | 41 |
| unit | **Vitamin A** | 1,110 | 2,170 |

*Based on year round average. For tomatoes marketed from November through May, value is about 12 mg. from June through October, 32g.
**Applies to product without calcium salts added.

# TOMATO JUICE

| | grams | Weight: | 243 | 182 |
|---|---|---|---|---|

| | | canned:<br>cup | canned:<br>glass<br>6 fl. oz. |
|---|---|---|---|
| % | **Water** | 94 | 94 |
| | **Calories** | 45 | 35 |
| | **Protein** | 2 | 2 |
| grams | **Fat** | Trace | Trace |
| | **Carbohydrate** | 10 | 8 |
| | **Calcium** | 17 | 13 |
| | **Phosphorus** | 44 | 33 |
| | **Iron** | 2.2 | 1.6 |
| | **Sodium** | 486 | 364 |
| milligrams | **Potassium** | 552 | 413 |
| | **Thiamin** | .12 | .09 |
| | **Riboflavin** | .07 | .05 |
| | **Niacin** | 1.9 | 1.5 |
| | **Ascorbic Acid** | 39 | 29 |
| unit | **Vitamin A** | 1,940 | 1,460 |

# TURNIP GREENS

| grams | **Weight:** | 145 | 165 |
|---|---|---|---|

| | | cooked, drained: from raw (leaves & stems) | cooked, drained: from frozen (chopped) |
|---|---|---|---|
| % | **Water** | 94 | 93 |
| | **Calories** | 30 | 40 |
| | **Protein** | 3 | 4 |
| grams { | **Fat** | Trace | Trace |
| | **Carbohydrate** | 5 | 6 |
| | **Calcium** | 252 | 195 |
| | **Phosphorus** | 49 | 64 |
| | **Iron** | 1.5 | 2.6 |
| milligrams { | **Sodium** | *N.A. | *28 |
| | **Potassium** | N.A. | 246 |
| | **Thiamin** | .15 | .08 |
| | **Riboflavin** | .33 | .15 |
| | **Niacin** | .7 | .7 |
| | **Ascorbic Acid** | 68 | 31 |
| unit | **Vitamin A** | 8,270 | 11,390 |

*Value for unsalted product.

# TURNIPS

grams **Weight:** 155

cooked, diced

| | | |
|---|---|---|
| % | **Water** | 94 |
| | **Calories** | 35 |
| grams | **Protein** | 1 |
| | **Fat** | Trace |
| | **Carbohydrate** | 8 |
| | **Calcium** | 54 |
| milligrams | **Phosphorus** | 37 |
| | **Iron** | .6 |
| | **Sodium** | *53 |
| | **Potassium** | 291 |
| | **Thiamin** | .06 |
| | **Riboflavin** | .08 |
| | **Niacin** | .5 |
| | **Ascorbic Acid** | 34 |
| unit | **Vitamin A** | Trace |

*Value is for unsalted product.

# SAUCES AND SALAD DRESSINGS

BARBEQUE SAUCE
BLUE CHEESE SALAD DRESSING
FRENCH SALAD DRESSING, COMMERCIAL
FRENCH SALAD DRESSING, HOMEMADE
ITALIAN SALAD DRESSING
MAYONNAISE
MAYONNAISE TYPE SALAD DRESSING
MUSTARD
RUSSIAN SALAD DRESSING
TARTAR SAUCE
THOUSAND ISLAND SALAD DRESSING
TOMATO CATSUP
WHITE SAUCE

# BARBECUE SAUCE

grams **Weight:** 250

| | | |
|---|---|---|
| % | **Water** | 81 |
| | **Calories** | 230 |
| grams | **Protein** | 4 |
| | **Fat** | 17 |
| | **Carbohydrate** | 20 |
| | **Calcium** | 53 |
| milligrams | **Phosphorus** | 50 |
| | **Iron** | 2.0 |
| | **Sodium** | 2,038 |
| | **Potassium** | 435 |
| | **Thiamin** | .03 |
| | **Riboflavin** | .03 |
| | **Niacin** | .8 |
| | **Ascorbic Acid** | 13 |
| unit | **Vitamin A** | 900 |

# BLUE CHEESE
## SALAD DRESSING

| grams | **Weight:** | 15 | 16 |
|---|---|---|---|
| | | regular | low calorie |

| | | | |
|---|---|---|---|
| % | **Water** | 32 | 84 |
| grams | **Calories** | 75 | 10 |
| | **Protein** | 1 | Trace |
| | **Fat** | 8 | 1 |
| | **Carbohydrate** | 1 | 1 |
| | **Calcium** | 12 | 10 |
| milligrams | **Phosphorus** | 11 | 8 |
| | **Iron** | Trace | Trace |
| | **Sodium** | 164 | 177 |
| | **Potassium** | 6 | 5 |
| | **Thiamin** | Trace | Trace |
| | **Riboflavin** | .02 | .01 |
| | **Niacin** | Trace | Trace |
| | **Ascorbic Acid** | Trace | Trace |
| unit | **Vitamin A** | 30 | 30 |

# FRENCH
## SALAD DRESSING

| | grams | Weight: | 16 | 16 |
|---|---|---|---|---|

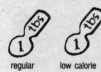

|  |  | regular | low calorie |
|---|---|---|---|

| | | regular | low calorie |
|---|---|---|---|
| % | **Water** | 39 | 77 |
| | **Calories** | 65 | 15 |
| grams | **Protein** | Trace | Trace |
| | **Fat** | 6 | 1 |
| | **Carbohydrate** | 3 | 2 |
| | **Calcium** | 2 | 2 |
| milligrams | **Phosphorus** | 2 | 2 |
| | **Iron** | .1 | .1 |
| | **Sodium** | 219 | 126 |
| | **Potassium** | 13 | 13 |
| | **Thiamin** | N.A. | N.A. |
| | **Riboflavin** | N.A. | N.A. |
| | **Niacin** | N.A. | N.A. |
| | **Ascorbic Acid** | N.A. | N.A. |
| unit | **Vitamin A** | N.A. | N.A. |

# FRENCH SALAD DRESSING
## (HOMEMADE)

grams | **Weight:** | 14

| | | |
|---|---|---|
| % | **Water** | 24.2 |
| | **Calories** | 88 |
| grams | **Protein** | Trace |
| | **Fat** | 9.8 |
| | **Carbohydrate** | .5 |
| | **Calcium** | 1 |
| | **Phosphorus** | Trace |
| | **Iron** | Trace |
| milligrams | **Sodium** | 92 |
| | **Potassium** | 4 |
| | **Thiamin** | N.A. |
| | **Riboflavin** | N.A. |
| | **Niacin** | N.A. |
| | **Ascorbic Acid** | N.A. |
| unit | **Vitamin A** | N.A. |

298

# SALAD DRESSING
# ITALIAN
# (COMMERCIAL)

| grams | **Weight:** | 15 | 15 |
|---|---|---|---|

| | | regular | low calorie |
|---|---|---|---|
| % | **Water** | 28 | 90 |
| | **Calories** | 85 | 10 |
| grams | **Protein** | Trace | Trace |
| | **Fat** | 9 | 1 |
| | **Carbohydrate** | 1 | Trace |
| | **Calcium** | 2 | Trace |
| milligrams | **Phosphorus** | 1 | 1 |
| | **Iron** | Trace | Trace |
| | **Sodium** | 314 | 118 |
| | **Potassium** | 2 | 2 |
| | **Thiamin** | Trace | Trace |
| | **Riboflavin** | Trace | Trace |
| | **Niacin** | Trace | Trace |
| | **Ascorbic Acid** | N.A. | N.A. |
| unit | **Vitamin A** | Trace | Trace |

# MAYONNAISE
## (COMMERCIAL)

grams  **Weight:**  14

| | | |
|---|---|---|
| % | **Water** | 15 |
| grams | **Calories** | 100 |
| | **Protein** | Trace |
| | **Fat** | 11 |
| | **Carbohydrate** | Trace |
| | **Calcium** | 3 |
| milligrams | **Phosphorus** | 4 |
| | **Iron** | .1 |
| | **Sodium** | 84 |
| | **Potassium** | 5 |
| | **Thiamin** | Trace – |
| | **Riboflavin** | .01 |
| | **Niacin** | Trace |
| | **Ascorbic Acid** | N.A. |
| unit | **Vitamin A** | 40 |

# SALAD DRESSING
# MAYONNAISE TYPE
## (COMMERCIAL)

| | | regular | low calorie |
|---|---|---|---|
| grams | **Weight:** | 15 | 16 |

| unit | | regular | low calorie |
|---|---|---|---|
| % | **Water** | 41 | 81 |
| | **Calories** | 65 | 20 |
| grams | **Protein** | Trace | Trace |
| | **Fat** | 6 | 2 |
| | **Carbohydrate** | 2 | 2 |
| | **Calcium** | 2 | 3 |
| milligrams | **Phosphorus** | 4 | 4 |
| | **Iron** | Trace | Trace |
| | **Sodium** | 88 | 19 |
| | **Potassium** | 1 | 1 |
| | **Thiamin** | Trace | Trace |
| | **Riboflavin** | Trace | Trace |
| | **Niacin** | Trace | Trace |
| | **Ascorbic Acid** | N.A. | N.A. |
| unit | **Vitamin A** | 30 | 40 |

301

# MUSTARD
## (PREPARED)

| grams | **Weight:** | 5 | 5 |
|---|---|---|---|

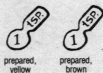

| | | prepared,<br>yellow | prepared,<br>brown |
|---|---|---|---|

| | | | |
|---|---|---|---|
| % | **Water** | 80 | 78.1 |
| | **Calories** | 5 | 5 |
| grams | **Protein** | Trace | .3 |
| | **Fat** | Trace | .3 |
| | **Carbohydrate** | Trace | .3 |
| | **Calcium** | 4 | 6 |
| milligrams | **Phosphorus** | 4 | 7 |
| | **Iron** | .1 | .1 |
| | **Sodium** | 63 | 65 |
| | **Potassium** | 7 | 7 |
| | **Thiamin** | N.A. | N.A. |
| | **Riboflavin** | N.A. | N.A. |
| | **Niacin** | N.A. | N.A. |
| | **Ascorbic Acid** | N.A. | N.A. |
| unit | **Vitamin A** | N.A. | N.A. |

# SALAD DRESSING
# RUSSIAN
## (COMMERCIAL)

grams     **Weight:**                    15

| | | |
|---|---|---|
| % | **Water** | 34.5 |
| | **Calories** | 74 |
| grams | **Protein** | .2 |
| | **Fat** | 1.6 |
| | **Carbohydrate** | 1.6 |
| | **Calcium** | 3 |
| milligrams | **Phosphorus** | 6 |
| | **Iron** | .1 |
| | **Sodium** | 130 |
| | **Potassium** | 24 |
| | **Thiamin** | .01 |
| | **Riboflavin** | .01 |
| | **Niacin** | .1 |
| | **Ascorbic Acid** | 1 |
| unit | **Vitamin A** | 100 |

# TARTAR SAUCE
## (COMMERCIAL)

| grams | Weight: | 14 | 14 |
|---|---|---|---|

| | regular | low calorie |
|---|---|---|

| % | Water | 34 | 68 |
|---|---|---|---|
| | **Calories** | 75 | 31 |
| grams | **Protein** | Trace | .1 |
| | **Fat** | 8 | 3 |
| | **Carbohydrate** | 1 | 1 |
| | **Calcium** | 3 | 3 |
| milligrams | **Phosphorus** | 4 | 4 |
| | **Iron** | .1 | .1 |
| | **Sodium** | 99 | 99 |
| | **Potassium** | 11 | 11 |
| | **Thiamin** | Trace | Trace |
| | **Riboflavin** | Trace | Trace |
| | **Niacin** | Trace | Trace |
| | **Ascorbic Acid** | Trace | Trace |
| unit | **Vitamin A** | 30 | 30 |

# THOUSAND ISLAND
# SALAD DRESSING
## (COMMERCIAL)

| grams | Weight: | 16 | 15 |
|---|---|---|---|

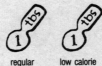

| | regular | low calorie |
|---|---|---|

| | | regular | low calorie |
|---|---|---|---|
| % | **Water** | 32 | 68 |
| | **Calories** | 80 | 25 |
| grams { | **Protein** | Trace | Trace |
| | **Fat** | 8 | 2 |
| | **Carbohydrate** | 2 | 2 |
| | **Calcium** | 2 | 2 |
| milligrams { | **Phosphorus** | 3 | 3 |
| | **Iron** | .1 | .1 |
| | **Sodium** | 112 | 105 |
| | **Potassium** | 18 | 17 |
| | **Thiamin** | Trace | Trace |
| | **Riboflavin** | Trace | Trace |
| | **Niacin** | Trace | Trace |
| | **Ascorbic Acid** | Trace | Trace |
| unit | **Vitamin A** | 50 | 50 |

# TOMATO CATSUP
## (COMMERCIAL, BOTTLED)

grams **Weight:** 273      15

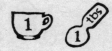

| | | |
|---|---|---|
| % | **Water** | 69 | 69 |
| | **Calories** | 290 | 15 |
| grams { | **Protein** | 5 | Trace |
| | **Fat** | 1 | Trace |
| | **Carbohydrate** | 69 | 4 |
| | **Calcium** | 60 | 3 |
| milligrams { | **Phosphorus** | 137 | 8 |
| | **Iron** | 2.2 | .1 |
| | **Sodium** | *2,845 | *156 |
| | **Potassium** | 991 | 54 |
| | **Thiamin** | .25 | .01 |
| | **Riboflavin** | .19 | .01 |
| | **Niacin** | 4.4 | .2 |
| | **Ascorbic Acid** | 41 | 2 |
| unit | **Vitamin A** | 3,820 | 210 |

*Applies to regular pack.

306

# WHITE SAUCE

grams **Weight:** 250

medium with
enriched flour

| | | |
|---|---|---|
| % | **Water** | 73 |
| grams { | **Calories** | 405 |
| | **Protein** | 10 |
| | **Fat** | 31 |
| | **Carbohydrate** | 22 |
| | **Calcium** | 288 |
| milligrams { | **Phosphorus** | 233 |
| | **Iron** | .5 |
| | **Sodium** | 948 |
| | **Potassium** | 348 |
| | **Thiamin** | .12 |
| | **Riboflavin** | .43 |
| | **Niacin** | .7 |
| | **Ascorbic Acid** | 2 |
| unit | **Vitamin A** | 1,150 |

# SOUPS

BEAN WITH PORK SOUP
BEEF BOUILLON
BEEF NOODLE SOUP
CHICKEN NOODLE SOUP
CLAM CHOWDER, MANHATTAN STYLE
CREAM OF CHICKEN SOUP
CREAM OF MUSHROOM SOUP
MINESTRONE SOUP
ONION SOUP
SPLIT PEA SOUP
TOMATO SOUP
TOMATO VEGETABLE SOUP WITH NOODLES
VEGETABLE BEEF SOUP
VEGETARIAN SOUP

# BEAN WITH PORK SOUP
## (CANNED, CONDENSED)

grams    **Weight:**    265

| | | |
|---|---|---|
| % | **Water** | *68.9 |
| | **Calories** | 38 |
| | **Protein** | 1.8 |
| grams | **Fat** | 1.3 |
| | **Carbohydrate** | 4.9 |
| | **Calcium** | 14 |
| | **Phosphorus** | 29 |
| | **Iron** | .5 |
| | **Sodium** | 229 |
| milligrams | **Potassium** | 90 |
| | **Thiamin** | .03 |
| | **Riboflavin** | .02 |
| | **Niacin** | .2 |
| | **Ascorbic Acid** | 1 |
| unit | **Vitamin A** | 150 |

*Prepared with equal volume of water.

# BEEF BOUILLON
## SOUP

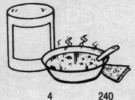

| grams | Weight: | 4 | 240 |
|---|---|---|---|
| | | bouillon cube, 1/3" | *canned, condensed |

| | | | |
|---|---|---|---|
| % | **Water** | 4 | 96 |
| | **Calories** | 5 | 30 |
| grams { | **Protein** | 1 | 5 |
| | **Fat** | Trace | 0 |
| | **Carbohydrate** | Trace | 3 |
| | **Calcium** | N.A. | Trace |
| | **Phosphorus** | N.A. | 31 |
| | **Iron** | N.A. | .5 |
| | **Sodium** | N.A. | 782 |
| milligrams { | **Potassium** | 4 | 130 |
| | **Thiamin** | N.A. | Trace |
| | **Riboflavin** | N.A. | .02 |
| | **Niacin** | N.A. | 1.2 |
| | **Ascorbic Acid** | N.A. | N.A. |
| unit | **Vitamin A** | N.A. | Trace |

*Prepared with equal volume of water.

# BEEF NOODLE SOUP

grams **Weight:** 240

*canned,
condensed

| | | |
|---|---|---|
| % | **Water** | 93 |
| | **Calories** | 65 |
| | **Protein** | 4 |
| grams | **Fat** | 3 |
| | **Carbohydrate** | 7 |
| | **Calcium** | 7 |
| | **Phosphorus** | 48 |
| | **Iron** | 1.0 |
| | **Sodium** | 917 |
| milligrams | **Potassium** | 77 |
| | **Thiamin** | .05 |
| | **Riboflavin** | .07 |
| | **Niacin** | 1.0 |
| | **Ascorbic Acid** | Trace |
| unit | **Vitamin A** | 50 |

*Prepared with equal volume of water.

# CHICKEN NOODLE
## SOUP

grams  **Weight:**  240

*dehydrated
mix

| | | |
|---|---|---|
| % | **Water** | 95 |
| | **Calories** | 55 |
| | **Protein** | 2 |
| | **Fat** | 1 |
| | **Carbohydrate** | 8 |
| | **Calcium** | 7 |
| | **Phosphorus** | 19 |
| | **Iron** | .2 |
| | **Sodium** | 979 |
| | **Potassium** | 19 |
| | **Thiamin** | .07 |
| | **Riboflavin** | .05 |
| | **Niacin** | .5 |
| | **Ascorbic Acid** | Trace |
| unit | **Vitamin A** | 50 |

grams { Calories, Protein, Fat, Carbohydrate, Calcium

milligrams { Phosphorus, Iron, Sodium, Potassium, Thiamin, Riboflavin, Niacin, Ascorbic Acid

*Prepared with equal volume of water.

312

# MANHATTAN
# CLAM CHOWDER
## (WITH TOMATOES, NO MILK)

grams **Weight:** 245

*canned,
condensed

| | | |
|---|---|---|
| % **Water** | 92 | |
| **Calories** | 80 | |
| grams { **Protein** | 2 | |
| **Fat** | 3 | |
| **Carbohydrate** | 12 | |
| **Calcium** | 34 | |
| **Phosphorus** | 47 | |
| **Iron** | 1.0 | |
| **Sodium** | 938 | |
| milligrams { **Potassium** | 184 | |
| **Thiamin** | .02 | |
| **Riboflavin** | .02 | |
| **Niacin** | 1.0 | |
| **Ascorbic Acid** | N.A. | |
| unit **Vitamin A** | 880 | |

*Prepared with equal volume of water.

# CREAM OF CHICKEN
## SOUP

| grams | **Weight:** | 240 | 245 |
|---|---|---|---|

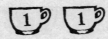

|  |  | *canned, condensed | **canned, condensed |
|---|---|---|---|
| % | **Water** | 92 | 85 |
| | **Calories** | 95 | 180 |
| grams | **Protein** | 3 | 7 |
| | **Fat** | 6 | 10 |
| | **Carbohydrate** | 8 | 15 |
| | **Calcium** | 24 | 172 |
| milligrams | **Phosphorus** | 34 | 152 |
| | **Iron** | .5 | .5 |
| | **Sodium** | 970 | 1,054 |
| | **Potassium** | 79 | 260 |
| | **Thiamin** | .02 | .05 |
| | **Riboflavin** | .05 | .27 |
| | **Niacin** | .5 | .7 |
| | **Ascorbic Acid** | Trace | 2 |
| unit | **Vitamin A** | 410 | 610 |

*Prepared with equal volume of water.
**Prepared with equal volume of milk.

# CREAM OF MUSHROOM
## SOUP

| | grams Weight: | 240 | 245 |
|---|---|---|---|

| | | *canned, condensed | **canned condensed |
|---|---|---|---|
| % | Water | 90 | 83 |
| | Calories | 135 | 215 |
| grams | Protein | 2 | 7 |
| | Fat | 10 | 14 |
| | Carbohydrate | 10 | 16 |
| | Calcium | 41 | 191 |
| milligrams | Phosphorus | 50 | 169 |
| | Iron | .5 | .5 |
| | Sodium | 955 | 1,039 |
| | Potassium | 98 | 279 |
| | Thiamin | .02 | .05 |
| | Riboflavin | .12 | .34 |
| | Niacin | .7 | .7 |
| | Ascorbic Acid | Trace | 1 |
| unit | Vitamin A | 70 | 250 |

*Prepared with equal volume of water.
**Prepared with equal volume of milk.

# MINESTRONE SOUP

| | grams | **Weight:** | 245 |

*canned,
condensed

| | | | |
|---|---|---|---|
| % | **Water** | 90 | |
| | **Calories** | 105 | |
| grams | **Protein** | 5 | |
| | **Fat** | 3 | |
| | **Carbohydrate** | 14 | |
| | **Calcium** | 37 | |
| | **Phosphorus** | 59 | |
| | **Iron** | 1.0 | |
| milligrams | **Sodium** | 995 | |
| | **Potassium** | 314 | |
| | **Thiamin** | .07 | |
| | **Riboflavin** | .05 | |
| | **Niacin** | 1.0 | |
| | **Ascorbic Acid** | N.A. | |
| unit | **Vitamin A** | 2,350 | |

*Prepared with equal volume of water.

# ONION SOUP

| | grams | **Weight:** | 43 | 240 |
|---|---|---|---|---|

| | | *dehydrated mix | **dehydrated mix |
|---|---|---|---|
| % | **Water** | 3 | 96 |
| | **Calories** | 150 | 35 |
| grams | **Protein** | 6 | 1 |
| | **Fat** | 5 | 1 |
| | **Carbohydrate** | 23 | 6 |
| | **Calcium** | 42 | 10 |
| milligrams | **Phosphorus** | 49 | 12 |
| | **Iron** | .6 | .2 |
| | **Sodium** | 2,871 | 689 |
| | **Potassium** | 238 | 58 |
| | **Thiamin** | .05 | Trace |
| | **Riboflavin** | .03 | .02 |
| | **Niacin** | .3 | .5 |
| | **Ascorbic Acid** | 6 | 5 |
| unit | **Vitamin A** | 30 | Trace |

*Unprepared.
**Prepared with equal volume of water.

# SPLIT PEA SOUP

| grams | **Weight:** | 245 |
|---|---|---|

*canned, condensed

| | | |
|---|---|---|
| % | **Water** | 85 |
| | **Calories** | 145 |
| grams { | **Protein** | 9 |
| | **Fat** | 3 |
| | **Carbohydrate** | 21 |
| | **Calcium** | 29 |
| milligrams { | **Phosphorus** | 149 |
| | **Iron** | 1.5 |
| | **Sodium** | 941 |
| | **Potassium** | 270 |
| | **Thiamin** | .25 |
| | **Riboflavin** | .15 |
| | **Niacin** | 1.5 |
| | **Ascorbic Acid** | 1 |
| unit | **Vitamin A** | 440 |

*Prepared with equal volume of water.

# TOMATO SOUP

| | grams | Weight: | 245 | 250 |
|---|---|---|---|---|

| | | *canned, condensed | **canned, condensed |
|---|---|---|---|
| % | **Water** | 91 | 84 |
| | **Calories** | 90 | 175 |
| | **Protein** | 2 | 7 |
| | **Fat** | 3 | 7 |
| | **Carbohydrate** | 16 | 23 |
| | **Calcium** | 15 | 168 |
| | **Phosphorus** | 34 | 155 |
| | **Iron** | .7 | .8 |
| | **Sodium** | 970 | 1,055 |
| | **Potassium** | 230 | 418 |
| | **Thiamin** | .05 | .10 |
| | **Riboflavin** | .05 | .25 |
| | **Niacin** | 1.2 | 1.3 |
| | **Ascorbic Acid** | 12 | 15 |
| unit | **Vitamin A** | 1,000 | 1,200 |

*Prepared with equal volume of water.
**Prepared with equal volume of milk.

# TOMATO VEGETABLE SOUP
## WITH NOODLES

grams    **Weight:**     240

\*dehydrated
mix

| | | |
|---|---|---|
| % | **Water** | 93 |
| | **Calories** | 65 |
| grams | **Protein** | 1 |
| | **Fat** | 1 |
| | **Carbohydrate** | 12 |
| | **Calcium** | 7 |
| milligrams | **Phosphorus** | 19 |
| | **Iron** | .2 |
| | **Sodium** | N.A. |
| | **Potassium** | 29 |
| | **Thiamin** | .05 |
| | **Riboflavin** | .02 |
| | **Niacin** | .5 |
| | **Ascorbic Acid** | 5 |
| unit | **Vitamin A** | 480 |

\*Prepared with equal volume of water.

# VEGETABLE BEEF
## SOUP

grams **Weight:** 245

*canned,
condensed

| | | |
|---|---|---|
| % | **Water** | 92 |
| | **Calories** | 80 |
| | **Protein** | 5 |
| | **Fat** | 2 |
| | **Carbohydrate** | 10 |
| | **Calcium** | 12 |
| | **Phosphorus** | 49 |
| | **Iron** | .7 |
| | **Sodium** | 1,046 |
| | **Potassium** | 162 |
| | **Thiamin** | .05 |
| | **Riboflavin** | .05 |
| | **Niacin** | 1.0 |
| | **Ascorbic Acid** | N.A. |
| unit | **Vitamin A** | 2,700 |

*grams* { Calories, Protein, Fat, Carbohydrate, Calcium }

*milligrams* { Phosphorus, Iron, Sodium, Potassium, Thiamin, Riboflavin, Niacin, Ascorbic Acid, Vitamin A }

*Prepared with equal volume of water.

# VEGETARIAN SOUP

grams **Weight:** 245

*canned,
condensed

| % | | |
|---|---|---|
| **Water** | 92 | |
| **Calories** | 80 | |
| **Protein** | 2 | |
| **Fat** | 2 | |
| **Carbohydrate** | 13 | |
| **Calcium** | 20 | |
| **Phosphorus** | 39 | |
| **Iron** | 1.0 | |
| **Sodium** | 838 | |
| **Potassium** | 172 | |
| **Thiamin** | .05 | |
| **Riboflavin** | .05 | |
| **Niacin** | 1.0 | |
| **Ascorbic Acid** | N.A. | |
| **Vitamin A** | 2,940 | |

grams { Calories–Calcium
milligrams { Phosphorus–Vitamin A
unit

*Prepared with equal volume of water.

322

# GRAINS AND GRAIN FLOURS

BARLEY

BUCKWHEAT FLOUR

BULGAR

CORNMEAL, WHOLE

CORNMEAL, DEGERMED

POPCORN

RICE, WHITE

RICE, BROWN

RYE FLOUR

WHOLE WHEAT FLOUR

WHEAT FLOUR, ENRICHED

SEE ALSO GRAIN PRODUCTS — BREADS, BREAKFAST CEREALS, PASTA, COOKIES, PIES, CAKES, ETC.

# BARLEY
## (PEARLED)

| | Weight: | 200 |
|---|---|---|
| grams | | |

uncooked

| | | |
|---|---|---|
| % | **Water** | 11 |
| | **Calories** | 700 |
| grams { | **Protein** | 16 |
| | **Fat** | 2 |
| | **Carbohydrate** | 158 |
| | **Calcium** | 32 |
| | **Phosphorus** | 378 |
| | **Iron** | 4.0 |
| milligrams { | **Sodium** | 6 |
| | **Potassium** | 320 |
| | **Thiamin** | .24 |
| | **Riboflavin** | .10 |
| | **Niacin** | 6.2 |
| | **Ascorbic Acid** | 0 |
| unit | **Vitamin A** | 0 |

# BUCKWHEAT
## FLOUR

grams  **Weight:**  98

light, sifted

| | | |
|---|---|---|
| % | Water | 12 |
| | Calories | 340 |
| grams { Protein | 6 |
| | Fat | 1 |
| | Carbohydrate | 78 |
| | Calcium | 11 |
| milligrams { Phosphorus | 86 |
| | Iron | 1.0 |
| | Sodium | N.A. |
| | Potassium | 314 |
| | Thiamin | .08 |
| | Riboflavin | .04 |
| | Niacin | .4 |
| | Ascorbic Acid | 0 |
| unit | Vitamin A | 0 |

# BULGAR
## (PARBOILED WHEAT)*

| | | Weight: | seasoned | dry |
|---|---|---|---|---|
| grams | | | 135 | 170 |

| | | | seasoned | dry |
|---|---|---|---|---|
| % | Water | | 56 | 10 |
| | Calories | | 245 | 602 |
| grams | Protein | | 8 | 19 |
| | Fat | | 4 | 2.6 |
| | Carbohydrate | | 44 | 128.7 |
| | Calcium | | 27 | 49 |
| milligrams | Phosphorus | | 263 | 575 |
| | Iron | | 1.9 | 6.3 |
| | Sodium | | 621 | N.A. |
| | Potassium | | 151 | 389 |
| | Thiamin | | .08 | .48 |
| | Riboflavin | | .05 | .24 |
| | Niacin | | 4.1 | 7.7 |
| | Ascorbic Acid | | 0 | 0 |
| unit | Vitamin A | | 0 | 0 |

*Made from hard red winter wheat, cooked (canned); uncooked (dry).

# CORNMEAL

| | | whole-ground, unbolted, dry form | bolted (nearly whole-grain) dry form |
|---|---|---|---|
| grams | **Weight:** | 122 | 122 |

| | | whole-ground, unbolted, dry form | bolted (nearly whole-grain) dry form |
|---|---|---|---|
| % | **Water** | 12 | 12 |
| | **Calories** | 435 | 440 |
| grams | **Protein** | 11 | 11 |
| | **Fat** | 5 | 4 |
| | **Carbohydrate** | 90 | 91 |
| | **Calcium** | 24 | 21 |
| milligrams | **Phosphorus** | 312 | 272 |
| | **Iron** | 2.9 | 2.2 |
| | **Sodium** | 1 | 1 |
| | **Potassium** | 346 | 303 |
| | **Thiamin** | .46 | .37 |
| | **Riboflavin** | .13 | .10 |
| | **Niacin** | 2.4 | 2.3 |
| | **Ascorbic Acid** | 0 | 0 |
| unit | **Vitamin A** | *620 | *590 |

*Applies to yellow varieties; white varieties contain only a trace.

# CORNMEAL
## (DEGERMED, COOKED)

| | grams | Weight: | 138 | 240 |
|---|---|---|---|---|

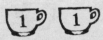

| | | enriched | unenriched |
|---|---|---|---|
| % | **Water** | 88 | 88 |
| | **Calories** | 120 | 120 |
| grams | **Protein** | 3 | 3 |
| | **Fat** | Trace | Trace |
| | **Carbohydrate** | 26 | 26 |
| | **Calcium** | 2 | 2 |
| milligrams | **Phosphorus** | 34 | 34 |
| | **Iron** | N.A. | 1.0 |
| | **Sodium** | **264 | **264 |
| | **Potassium** | 38 | 38 |
| | **Thiamin** | .14 | .05 |
| | **Riboflavin** | .10 | .02 |
| | **Niacin** | 1.2 | .2 |
| | **Ascorbic Acid** | 0 | 0 |
| unit | **Vitamin A** | *140 | *140 |

*Applies to yellow varieties; white varieties contain only a trace.
**Based on value of 110 mg. per 100g for product cooked with salt added as specified by manufacturers. If cooked without added salt, value is negligible.

# CORNMEAL
## (DEGERMED, DRY)

| | grams | Weight: | 138 | 138 |
|---|---|---|---|---|

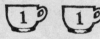

|  | enriched | unenriched |
|---|---|---|

| | | enriched | unenriched |
|---|---|---|---|
| % | Water | 12 | 12 |
| | Calories | 500 | 500 |
| grams | Protein | 11 | 11 |
| | Fat | 2 | 2 |
| | Carbohydrate | 108 | 108 |
| | Calcium | 8 | 8 |
| milligrams | Phosphorus | 137 | 137 |
| | Iron | 4.0 | 1.5 |
| | Sodium | 1 | 1 |
| | Potassium | 166 | 166 |
| | Thiamin | .61 | .19 |
| | Riboflavin | .36 | .07 |
| | Niacin | 4.8 | 1.4 |
| | Ascorbic Acid | 0 | 0 |
| unit | Vitamin A | *610 | *610 |

*Applies to yellow varieties; white varieties contain only a trace.

# POPCORN

| | grams | Weight: | 6 | 9 | 35 |
|---|---|---|---|---|---|

| | | | plain, large kernel | with oil (coconut) & salt added, large kernel | sugar coated |
|---|---|---|---|---|---|
| % | Water | | 4 | 3 | 4 |
| | Calories | | 25 | 40 | 135 |
| grams | Protein | | 1 | 1 | 2 |
| | Fat | | Trace | 2 | 1 |
| | Carbohydrate | | 5 | 5 | 30 |
| | Calcium | | 1 | 1 | 2 |
| milligrams | Phosphorus | | 17 | 19 | 47 |
| | Iron | | .2 | .2 | .5 |
| | Sodium | | Trace | 175 | *Trace |
| | Potassium | | N.A. | N.A. | N.A. |
| | Thiamin | | N.A. | N.A. | N.A. |
| | Riboflavin | | .01 | .01 | .02 |
| | Niacin | | .1 | .2 | .4 |
| | Ascorbic Acid | | 0 | 0 | 0 |
| unit | Vitamin A | | N.A. | N.A. | N.A. |

*Value for product without added salt.

330

# RICE, WHITE
## (ENRICHED)

| | grams Weight: | 165 | 185 | 205 |
|---|---|---|---|---|
| | | instant, ready-to-serve, hot | long grain: raw | long grain: cooked, served hot |
| % | Water | 73 | 12 | 13 |
| | Calories | 180 | 670 | 225 |
| grams | Protein | 4 | 12 | 4 |
| | Fat | Trace | 1 | Trace |
| | Carbohydrate | 40 | 149 | 50 |
| | Calcium | 5 | 44 | 21 |
| milligrams | Phosphorus | 31 | 174 | 57 |
| | Iron | 1.3 | 5.4 | 1.8 |
| | Sodium | **450 | 9 | **767 |
| | Potassium | N.A. | 170 | 57 |
| | Thiamin | .21 | .81 | .23 |
| | Riboflavin | * | .06 | .02 |
| | Niacin | 1.7 | 6.5 | 2.1 |
| | Ascorbic Acid | 0 | 0 | 0 |
| unit | Vitamin A | 0 | 0 | 0 |

*Product may or may not be enriched with riboflavin, consult label.
**Applies to product cooked with salt added as specified by manufacturers. If cooked without added salt, value is negligible.

# RICE, BROWN

| | grams | Weight: | 185 | 195 |
|---|---|---|---|---|

| | | long grain, raw | long grain, cooked, hot |
|---|---|---|---|

| | | | |
|---|---|---|---|
| % | **Water** | 12 | 70.3 |
| | **Calories** | 666 | 232 |
| grams | **Protein** | 13.9 | 4.9 |
| | **Fat** | 3.5 | 1.2 |
| | **Carbohydrate** | 143.2 | 49.7 |
| | **Calcium** | 59 | 23 |
| milligrams | **Phosphorus** | 409 | 142 |
| | **Iron** | 3 | 1 |
| | **Sodium** | 17 | *550 |
| | **Potassium** | 396 | 137 |
| | **Thiamin** | .63 | .18 |
| | **Riboflavin** | .09 | .04 |
| | **Niacin** | 8.7 | 2.7 |
| | **Ascorbic Acid** | 0 | 0 |
| unit | **Vitamin A** | 0 | 0 |

*Cooked with salt. If unsalted, sodium value is negligible.

# RYE FLOUR

| | grams | Weight: | 102 | 88 |
|---|---|---|---|---|

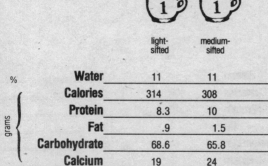

| | | light-<br>sifted | medium-<br>sifted |
|---|---|---|---|
| % | Water | 11 | 11 |
| | Calories | 314 | 308 |
| grams | Protein | 8.3 | 10 |
| | Fat | .9 | 1.5 |
| | Carbohydrate | 68.6 | 65.8 |
| | Calcium | 19 | 24 |
| milligrams | Phosphorus | 163 | 231 |
| | Iron | 1 | 2.3 |
| | Sodium | 1 | 1 |
| | Potassium | 137 | 179 |
| | Thiamin | .13 | .26 |
| | Riboflavin | .06 | .11 |
| | Niacin | .5 | 2.2 |
| | Ascorbic Acid | 0 | 0 |
| unit | Vitamin A | 0 | 0 |

# WHOLE WHEAT
## FLOUR

grams **Weight:** 120

| | from hard-wheats, stirred |
|---|---|
| % | |
| **Water** | 12 |
| **Calories** | 400 |
| **Protein** | 16 |
| **Fat** | 2 |
| **Carbohydrate** | 85 |
| **Calcium** | 49 |
| **Phosphorus** | 446 |
| **Iron** | 4.0 |
| **Sodium** | 4 |
| **Potassium** | 444 |
| **Thiamin** | .66 |
| **Riboflavin** | .14 |
| **Niacin** | 5.2 |
| **Ascorbic Acid** | 0 |
| **Vitamin A** | 0 |

grams { Protein, Fat, Carbohydrate, Calcium

milligrams { Phosphorus ... Ascorbic Acid

unit Vitamin A

# WHEAT FLOUR
## (WHITE, ENRICHED)

| | grams | Weight: | 115 | 125 |
|---|---|---|---|---|

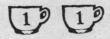

| | | all-purpose enriched; sifted, spooned | all-purpose enriched; unsifted, spooned |
|---|---|---|---|
| % | Water | 12 | 12 |
| | Calories | 420 | 455 |
| grams | Protein | 12 | 13.1 |
| | Fat | 1 | 1.3 |
| | Carbohydrate | 88 | 95.1 |
| | Calcium | 18 | 20 |
| milligrams | Phosphorus | 100 | 109 |
| | Iron | *3.3 | *3.6 |
| | Sodium | 2 | 3 |
| | Potassium | 109 | 119 |
| | Thiamin | .74 | *.55 |
| | Riboflavin | .46 | *.33 |
| | Niacin | 6.1 | *4.4 |
| | Ascorbic Acid | 0 | 0 |
| unit | Vitamin A | 0 | 0 |

*Based on product with minimum level of enrichment.

335

# WHEAT FLOUR
## (WHITE, ENRICHED)

| | grams | Weight: | 96 | 125 |
|---|---|---|---|---|

| | | cake or pastry flour enriched, spooned, sifted | self-rising enriched unsifted, spooned |
|---|---|---|---|
| % | Water | 12 | 12 |
| | Calories | 350 | 440 |
| grams | Protein | 7 | 12 |
| | Fat | 1 | 1 |
| | Carbohydrate | 76 | 93 |
| | Calcium | 16 | 331 |
| milligrams | Phosphorus | 70 | 583 |
| | Iron | 2.8 | 3.6 |
| | Sodium | 2 | 1,339 |
| | Potassium | 70 | N.A. |
| | Thiamin | .61 | .80 |
| | Riboflavin | .38 | .50 |
| | Niacin | 5.1 | 6.6 |
| | Ascorbic Acid | 0 | 0 |
| unit | Vitamin A | 0 | 0 |

# BREADS, ROLLS AND CRACKERS

BAGELS
BISCUITS
BOSTON BROWN BREAD
BREAD CRUMBS
BREAD STUFFING
CRACKED WHEAT BREAD
CRACKERS
FRENCH OR VIENNA BREAD
ITALIAN BREAD
MUFFINS, HOMEMADE
MUFFINS, FROM MIX
PIZZA, CHEESE
PRETZELS
PUMPERNICKEL BREAD
RAISIN BREAD
ROLLS, COMMERCIAL
ROLLS, HOMEMADE
RYE BREAD
WHITE BREAD (FIRM CRUMB), 1 LB. LOAF
AND SLICES
WHITE BREAD (FIRM CRUMB), 2 LB. LOAF
AND SLICES

WHITE BREAD (SOFT CRUMB), 1 LB. LOAF
AND SLICES

WHITE BREAD (SOFT CRUMB), 1½ LB. LOAF
AND SLICES

WHITE BREAD (SOFT CRUMB), CUBES AND
CRUMBS

WHOLE WHEAT BREAD (FIRM CRUMB)

WHOLE WHEAT BREAD (SOFT CRUMB)

# BAGELS

| grams | **Weight:** | 55 | 55 |
|---|---|---|---|

| | | egg | water |
|---|---|---|---|

| | | | |
|---|---|---|---|
| % | **Water** | 32 | 29 |
| | **Calories** | 165 | 165 |
| grams | **Protein** | 6 | 6 |
| | **Fat** | 2 | 1 |
| | **Carbohydrate** | 28 | 30 |
| | **Calcium** | 9 | 8 |
| milligrams | **Phosphorus** | 43 | 41 |
| | **Iron** | 1.2 | 1.2 |
| | **Sodium** | N.A. | N.A. |
| | **Potassium** | 41 | 42 |
| | **Thiamin** | .14 | .15 |
| | **Riboflavin** | .10 | .11 |
| | **Niacin** | 1.2 | 1.4 |
| | **Ascorbic Acid** | 0 | 0 |
| unit | **Vitamin A** | 30 | 0 |

# BISCUITS

| | grams | **Weight:** | 28 | 28 |
|---|---|---|---|---|

| | | from home recipe* | from mix** |
|---|---|---|---|
| % | **Water** | 27 | 29 |
| | **Calories** | 105 | 90 |
| grams | **Protein** | 2 | 2 |
| | **Fat** | 5 | 3 |
| | **Carbohydrate** | 13 | 15 |
| | **Calcium** | 34 | 19 |
| milligrams | **Phosphorus** | 49 | 65 |
| | **Iron** | .4 | .6 |
| | **Sodium** | 175 | 272 |
| | **Potassium** | 33 | 32 |
| | **Thiamin** | .08 | .09 |
| | **Riboflavin** | .08 | .08 |
| | **Niacin** | .7 | .8 |
| | **Ascorbic Acid** | Trace | Trace |
| unit | **Vitamin A** | Trace | Trace |

*Enriched flour, vegetable shortening.
**Enriched flour, vegetable shortening, milk.
(2 inch diameter)

# BOSTON BROWN BREAD

grams    **Weight:**    45

canned, slice
3¼ by ½"*

| | | |
|---|---|---|
| % | **Water** | 45 |
| | **Calories** | 95 |
| grams { | **Protein** | 2 |
| | **Fat** | 1 |
| | **Carbohydrate** | 21 |
| | **Calcium** | 41 |
| milligrams { | **Phosphorus** | 72 |
| | **Iron** | .9 |
| | **Sodium** | 113 |
| | **Potassium** | 131 |
| | **Thiamin** | .06 |
| | **Riboflavin** | .04 |
| | **Niacin** | .7 |
| | **Ascorbic Acid** | 0 |
| unit | **Vitamin A** | **0 |

*Made with vegetable shortening.
**Applies to products made with white cornmeal. With yellow cornmeal, value is
30 International Units (I.U.).

341

# BREAD CRUMBS
## ENRICHED*

| | grams | **Weight:** | 45 |

dry grated

| | | |
|---|---|---|
| % | **Water** | 35 |
| | **Calories** | 124 |
| grams | **Protein** | 4.1 |
| | **Fat** | 1.7 |
| | **Carbohydrate** | 22.6 |
| | **Calcium** | 43 |
| | **Phosphorus** | 46 |
| | **Iron** | 1.1 |
| | **Sodium** | 223 |
| milligrams | **Potassium** | 54 |
| | **Thiamin** | .12 |
| | **Riboflavin** | .09 |
| | **Niacin** | 1.1 |
| | **Ascorbic Acid** | Trace |
| unit | **Vitamin A** | Trace |

*Made with vegetable shortening.

# BREAD STUFFING
## FROM MIX

| | | | |
|---|---|---:|---:|
| grams | **Weight:** | 140 | 200 |

| | | dry crumbly type* | moist type** |
|---|---|---:|---:|
| % | **Water** | 33.2 | 61.4 |
| grams | **Calories** | 501 | 416 |
| | **Protein** | 9.1 | 8.8 |
| | **Fat** | 30.5 | 25.6 |
| | **Carbohydrate** | 49.8 | 39.4 |
| | **Calcium** | 92 | 80 |
| milligrams | **Phosphorus** | 136 | 132 |
| | **Iron** | 2.2 | 2 |
| | **Sodium** | 1,254 | 1,008 |
| | **Potassium** | 126 | 116 |
| | **Thiamin** | .13 | .10 |
| | **Riboflavin** | .17 | .18 |
| | **Niacin** | 2.1 | 1.6 |
| | **Ascorbic Acid** | Trace | Trace |
| unit | **Vitamin A** | 910 | 840 |

*Prepared with water and table fat.
**Prepared with water, egg and table fat.

# CRACKED WHEAT BREAD*

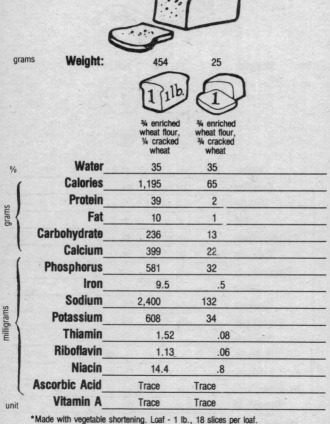

| | grams | Weight: | 454 | 25 |
|---|---|---|---|---|
| | | | ¾ enriched wheat flour, ¼ cracked wheat | ¾ enriched wheat flour, ¾ cracked wheat |
| % | | Water | 35 | 35 |
| grams { | | Calories | 1,195 | 65 |
| | | Protein | 39 | 2 |
| | | Fat | 10 | 1 |
| | | Carbohydrate | 236 | 13 |
| | | Calcium | 399 | 22 |
| milligrams { | | Phosphorus | 581 | 32 |
| | | Iron | 9.5 | .5 |
| | | Sodium | 2,400 | 132 |
| | | Potassium | 608 | 34 |
| | | Thiamin | 1.52 | .08 |
| | | Riboflavin | 1.13 | .06 |
| | | Niacin | 14.4 | .8 |
| | | Ascorbic Acid | Trace | Trace |
| unit | | Vitamin A | Trace | Trace |

*Made with vegetable shortening. Loaf - 1 lb., 18 slices per loaf.

# CRACKERS*

| | | graham, plain** | rye wafers, whole grain*** |
|---|---|---|---|
| grams | Weight: | 14 | 13 |
| % | Water | 6 | 6 |
| | Calories | 55 | 45 |
| grams | Protein | 1 | 2 |
| | Fat | 1 | Trace |
| | Carbohydrate | 10 | 10 |
| | Calcium | 6 | 7 |
| milligrams | Phosphorus | 21 | 50 |
| | Iron | .5 | .5 |
| | Sodium | 95 | N.A. |
| | Potassium | 55 | 78 |
| | Thiamin | .02 | .04 |
| | Riboflavin | .08 | .03 |
| | Niacin | .5 | .2 |
| | Ascorbic Acid | 0 | 0 |
| unit | Vitamin A | 0 | 0 |

*Made with vegetable shortening.
**2½" square
***1⅞" by 3½"

# CRACKERS*

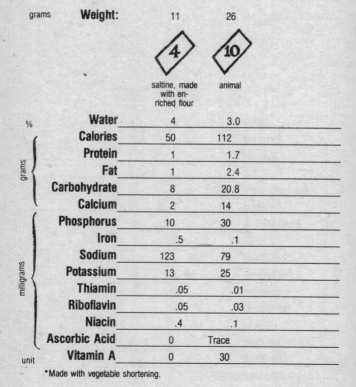

| | grams Weight: | 11 | 26 |
|---|---|---|---|
| | | **4** | **10** |
| | | saltine, made with en-riched flour | animal |
| % | **Water** | 4 | 3.0 |
| grams | **Calories** | 50 | 112 |
| | **Protein** | 1 | 1.7 |
| | **Fat** | 1 | 2.4 |
| | **Carbohydrate** | 8 | 20.8 |
| | **Calcium** | 2 | 14 |
| milligrams | **Phosphorus** | 10 | 30 |
| | **Iron** | .5 | .1 |
| | **Sodium** | 123 | 79 |
| | **Potassium** | 13 | 25 |
| | **Thiamin** | .05 | .01 |
| | **Riboflavin** | .05 | .03 |
| | **Niacin** | .4 | .1 |
| | **Ascorbic Acid** | 0 | Trace |
| unit | **Vitamin A** | 0 | 30 |

*Made with vegetable shortening.

# FRENCH OR VIENNA
# BREADS*

| | | | |
|---|---|---|---|
| grams | **Weight:** | 454 | 35 | 25 |

| | | loaf, 1 lb. | slice (5 by 2½ by 2"): French | slice (4¾ by 4 by ½"): Vienna |
|---|---|---|---|---|
| % | **Water** | 31 | 31 | 31 |
| | **Calories** | 1,315 | 100 | 75 |
| | **Protein** | 41 | 3 | 2 |
| grams | **Fat** | 14 | 1 | 1 |
| | **Carbohydrate** | 251 | 19 | 14 |
| | **Calcium** | 195 | 15 | 11 |
| | **Phosphorus** | 386 | 30 | 21 |
| | **Iron** | 10.0 | .8 | .6 |
| milligrams | **Sodium** | 2,631 | 203 | 145 |
| | **Potassium** | 408 | 32 | 23 |
| | **Thiamin** | 1.80 | .14 | .10 |
| | **Riboflavin** | 1.10 | .08 | .06 |
| | **Niacin** | 15.0 | 1.2 | .8 |
| | **Ascorbic Acid** | Trace | Trace | Trace |
| unit | **Vitamin A** | Trace | Trace | Trace |

*Made with vegetable shortening.

# ITALIAN BREAD

| | enriched loaf 1 lb. | enriched slice 4½ by 3¼ by ¾" |
|---|---|---|
| grams **Weight:** | 454 | 30 |
| % **Water** | 32 | 32 |
| **Calories** | 1,250 | 85 |
| **Protein** | 41 | 3 |
| **Fat** | 4 | Trace |
| **Carbohydrate** | 256 | 17 |
| **Calcium** | 77 | 5 |
| **Phosphorus** | 349 | 23 |
| **Iron** | 10.0 | .7 |
| **Sodium** | 2,654 | 176 |
| **Potassium** | 336 | 22 |
| **Thiamin** | 1.80 | .12 |
| **Riboflavin** | 1.10 | .07 |
| **Niacin** | 15.0 | 1.0 |
| **Ascorbic Acid** | 0 | 0 |
| unit **Vitamin A** | 0 | 0 |

grams { Calories, Protein, Fat, Carbohydrate, Calcium

milligrams { Phosphorus, Iron, Sodium, Potassium, Thiamin, Riboflavin, Niacin, Ascorbic Acid

# MUFFINS
## HOME RECIPE

| | grams | **Weight:** | 40 | 40 |
|---|---|---|---|---|

| | | blueberry,*<br>2⅜″ diam.,<br>1½″ high | bran |
|---|---|---|---|

| | | | |
|---|---|---:|---:|
| % | **Water** | 39 | 35 |
| | **Calories** | 110 | 105 |
| grams | **Protein** | 3 | 3 |
| | **Fat** | 4 | 4 |
| | **Carbohydrate** | 17 | 17 |
| | **Calcium** | 34 | 57 |
| milligrams | **Phosphorus** | 53 | 162 |
| | **Iron** | .6 | 1.5 |
| | **Sodium** | 253 | 179 |
| | **Potassium** | 46 | 172 |
| | **Thiamin** | .09 | .07 |
| | **Riboflavin** | .10 | .10 |
| | **Niacin** | .7 | 1.7 |
| | **Ascorbic Acid** | Trace | Trace |
| unit | **Vitamin A** | 90 | 90 |

*Made with enriched flour.

# MUFFINS
## HOME RECIPE

| | grams | Weight: | 40 | 40 |
|---|---|---|---|---|

| | | corn* | plain<br>3" diam.<br>1½" high |
|---|---|---|---|
| % | Water | 33 | 38 |
| | Calories | 125 | 120 |
| grams | Protein | 3 | 3 |
| | Fat | 4 | 4 |
| | Carbohydrate | 19 | 17 |
| | Calcium | 42 | 42 |
| milligrams | Phosphorus | 68 | 60 |
| | Iron | .7 | .6 |
| | Sodium | 192 | 176 |
| | Potassium | 54 | 50 |
| | Thiamin | .10 | .09 |
| | Riboflavin | .10 | .12 |
| | Niacin | .7 | .9 |
| | Ascorbic Acid | Trace | Trace |
| unit | Vitamin A | **120 | 40 |

*Enriched degermed cornmeal and flour, 2⅜" diam., 1½" high.
**Applies to product made with yellow cornmeal.

350

# CORN MUFFINS
## (FROM MIX)*

grams  **Weight:**  40

1

2⅜″ diam.,
1½″ high*

| | | |
|---|---|---|
| % | **Water** | 30 |
| | **Calories** | 130 |
| grams | **Protein** | 3 |
| | **Fat** | 4 |
| | **Carbohydrate** | 20 |
| | **Calcium** | 96 |
| | **Phosphorus** | 150 |
| | **Iron** | .6 |
| milligrams | **Sodium** | 192 |
| | **Potassium** | 44 |
| | **Thiamin** | .08 |
| | **Riboflavin** | .09 |
| | **Niacin** | .7 |
| | **Ascorbic Acid** | Trace |
| unit | **Vitamin A** | **100 |

*Made with enriched degermed cornmeal and enriched flour, egg & milk.
**Applies to product made with yellow cornmeal.

# PIZZA, CHEESE*

| | | |
|---|---|---|
| grams | **Weight:** | 60 |

4¾″ sector;
⅛ of 12″
diam. pie*

| | | |
|---|---|---|
| % | **Water** | 45 |
| | **Calories** | 145 |
| grams | **Protein** | 6 |
| | **Fat** | 4 |
| | **Carbohydrate** | 22 |
| | **Calcium** | 86 |
| milligrams | **Phosphorus** | 89 |
| | **Iron** | 1.1 |
| | **Sodium** | 380 |
| | **Potassium** | 67 |
| | **Thiamin** | .16 |
| | **Riboflavin** | .18 |
| | **Niacin** | 1.6 |
| | **Ascorbic Acid** | 4 |
| unit | **Vitamin A** | 230 |

*Crust made with vegetable shortening and enriched flour.

# PRETZELS*

| grams | **Weight:** | 16 | 60 | 3 |
|---|---|---|---|---|

| | | Dutch twisted, 2¾ by 2⅝" | thin, twisted, 3¼ by 2¼ by ¼" | stick, 2¼" long |
|---|---|---|---|---|
| % | **Water** | 5 | 5 | 5 |
| | **Calories** | 60 | 235 | 10 |
| | **Protein** | 2 | 6 | Trace |
| grams | **Fat** | 1 | 3 | Trace |
| | **Carbohydrate** | 12 | 46 | 2 |
| | **Calcium** | 4 | 13 | 1 |
| | **Phosphorus** | 21 | 79 | 4 |
| | **Iron** | .2 | .9 | Trace |
| | **Sodium** | **269 | **1,008 | **50 |
| milligrams | **Potassium** | 21 | 78 | 4 |
| | **Thiamin** | .05 | .20 | .01 |
| | **Riboflavin** | .04 | .15 | .01 |
| | **Niacin** | .7 | 2.5 | .1 |
| | **Ascorbic Acid** | 0 | 0 | 0 |
| unit | **Vitamin A** | 0 | 0 | 0 |

*Made with enriched flour.
**Sodium content is variable. For example, very thin pretzel sticks contain about twice the average amount listed.

# PUMPERNICKEL
## BREAD

| | grams | Weight: | 454 | 32 |
|---|---|---|---|---|

* *

| | | 454 | 32 |
|---|---|---|---|
| % | **Water** | 34 | 34 |
| grams | **Calories** | 1,115 | 80 |
| | **Protein** | 41 | 3 |
| | **Fat** | 5 | Trace |
| | **Carbohydrate** | 241 | 17 |
| milligrams | **Calcium** | 381 | 27 |
| | **Phosphorus** | 1,039 | 73 |
| | **Iron** | 11.8 | .8 |
| | **Sodium** | 2,581 | 182 |
| | **Potassium** | 2,059 | 145 |
| | **Thiamin** | 1.30 | .09 |
| | **Riboflavin** | .93 | .07 |
| | **Niacin** | 8.5 | .6 |
| | **Ascorbic Acid** | 0 | 0 |
| unit | **Vitamin A** | 0 | 0 |

*⅔ rye flour, ⅓ enriched wheat flour.

# RAISIN BREAD
## ENRICHED

| grams | **Weight:** | 454 | 25 |
|---|---|---|---|

| | | loaf, 1 lb.* | slice<br>(18 per loaf) |
|---|---|---|---|

| | | | |
|---|---|---|---|
| % | **Water** | 35 | 35 |
| | **Calories** | 1,190 | 65 |
| grams | **Protein** | 30 | 2 |
| | **Fat** | 13 | 1 |
| | **Carbohydrate** | 243 | 13 |
| | **Calcium** | 322 | 18 |
| milligrams | **Phosphorus** | 395 | 22 |
| | **Iron** | 10.0 | .6 |
| | **Sodium** | 1,656 | 91 |
| | **Potassium** | 1,057 | 58 |
| | **Thiamin** | 1.70 | .09 |
| | **Riboflavin** | 1.07 | .06 |
| | **Niacin** | 10.7 | .6 |
| | **Ascorbic Acid** | Trace | Trace |
| unit | **Vitamin A** | Trace | Trace |

*Made with vegetable shortening.

# ROLLS
## COMMERCIAL*

| | grams | Weight: | 26 | 28 | 40 |
|---|---|---|---|---|---|
| | | | brown & serve (12 per 12 oz. pkg.) browned | cloverleaf or pan, 2½" diam., 2" high | frankfurter and hamburger |
| % | | **Water** | 27 | 31 | 31 |
| grams | | **Calories** | 85 | 85 | 120 |
| | | **Protein** | 2 | 2 | 3 |
| | | **Fat** | 2 | 2 | 2 |
| | | **Carbohydrate** | 14 | 15 | 21 |
| milligrams | | **Calcium** | 20 | 21 | 30 |
| | | **Phosphorus** | 23 | 24 | 34 |
| | | **Iron** | .5 | .5 | .8 |
| | | **Sodium** | 136 | 142 | 202 |
| | | **Potassium** | 25 | 27 | 38 |
| | | **Thiamin** | .10 | .11 | .16 |
| | | **Riboflavin** | .06 | .07 | .10 |
| | | **Niacin** | .9 | .9 | 1.3 |
| | | **Ascorbic Acid** | Trace | Trace | Trace |
| unit | | **Vitamin A** | Trace | Trace | Trace |

*Enriched, made with vegetable shortening.

# ROLLS
## COMMERCIAL*

| | grams | **Weight:** | 50 | 135 |
|---|---|---|---|---|

| | hard, 3¾"<br>diam., 2" | hoggie or<br>submarine<br>11½ x 3 x 2½" |
|---|---|---|

| | | | |
|---|---|---|---|
| % | **Water** | 25 | 31 |
| | **Calories** | 155 | 390 |
| grams { | **Protein** | 5 | 12 |
| | **Fat** | 2 | 4 |
| | **Carbohydrate** | 30 | 75 |
| | **Calcium** | 24 | 58 |
| milligrams { | **Phosphorus** | 46 | 115 |
| | **Iron** | 1.2 | 3.0 |
| | **Sodium** | 313 | N.A. |
| | **Potassium** | 49 | 122 |
| | **Thiamin** | .20 | .54 |
| | **Riboflavin** | .12 | .32 |
| | **Niacin** | 1.7 | 4.5 |
| | **Ascorbic Acid** | Trace | Trace |
| unit | **Vitamin A** | Trace | Trace |

*Enriched, made with vegetable shortening.

# ROLLS
## HOMEMADE*

grams **Weight:** 35

cloverleaf,
2½" diam.,
2" high

| % | Water | 26 |
|---|---|---|
| | **Calories** | 120 |
| grams | Protein | 3 |
| | Fat | 3 |
| | Carbohydrate | 20 |
| | Calcium | 16 |
| milligrams | Phosphorus | 36 |
| | Iron | .7 |
| | Sodium | .98 |
| | Potassium | 41 |
| | Thiamin | .12 |
| | Riboflavin | .12 |
| | Niacin | 1.2 |
| | Ascorbic Acid | Trace |
| unit | Vitamin A | 30 |

*Made with enriched flour.

# RYE BREAD
## LIGHT*

| | grams | Weight: | 454 | 25 |
|---|---|---|---|---|

loaf 1 lb.

| | | | |
|---|---|---|---|
| % | **Water** | 36 | 36 |
| | **Calories** | 1,100 | 60 |
| grams | **Protein** | 41 | 2 |
| | **Fat** | 5 | Trace |
| | **Carbohydrate** | 236 | 13 |
| | **Calcium** | 340 | 19 |
| | **Phosphorus** | 667 | 37 |
| milligrams | **Iron** | 9.1 | .5 |
| | **Sodium** | 2,527 | 139 |
| | **Potassium** | 658 | 36 |
| | **Thiamin** | 1.35 | .07 |
| | **Riboflavin** | .98 | .05 |
| | **Niacin** | 12.9 | .7 |
| | **Ascorbic Acid** | 0 | 0 |
| unit | **Vitamin A** | 0 | 0 |

*Made with ⅔ enriched wheat flour and ⅓ rye flour.

# ENRICHED*
# WHITE BREAD
## (FIRM CRUMB TYPE)

| grams | Weight: | 454 | 23 | 20 |
|---|---|---|---|---|
| | | loaf 1 lb. | slice (20 per loaf) | slice (20 per loaf) toasted |
| % | Water | 35 | 35 | 24 |
| | Calories | 1,245 | 65 | 65 |
| grams | Protein | 41 | 2 | 2 |
| | Fat | 17 | 1 | 1 |
| | Carbohydrate | 228 | 12 | 12 |
| | Calcium | 435 | 22 | 22 |
| milligrams | Phosphorus | 463 | 23 | 23 |
| | Iron | 11.3 | .6 | .6 |
| | Sodium | 2,245 | 114 | 114 |
| | Potassium | 549 | 28 | 28 |
| | Thiamin | 1.80 | .09 | .07 |
| | Riboflavin | 1.10 | .06 | .06 |
| | Niacin | 15.0 | .8 | .8 |
| | Ascorbic Acid | Trace | Trace | Trace |
| unit | Vitamin A | Trace | Trace | Trace |

*Made with vegetable shortening.

# ENRICHED*
# WHITE BREAD
## (FIRM CRUMB TYPE)

| grams | **Weight:** | 907 | 27 | 23 |
|---|---|---|---|---|
| | | loaf 2 lb. | slice (34 per loaf) | slice (34 per loaf) toasted |

| | | | | |
|---|---|---|---|---|
| % | **Water** | 35 | 35 | 24 |
| | **Calories** | 2,495 | 75 | 75 |
| grams { | **Protein** | 82 | 2 | 2 |
| | **Fat** | 34 | 1 | 1 |
| | **Carbohydrate** | 455 | 14 | 14 |
| | **Calcium** | 871 | 26 | 26 |
| milligrams { | **Phosphorus** | 925 | 28 | 28 |
| | **Iron** | 22.7 | .7 | .7 |
| | **Sodium** | 4,490 | 134 | 134 |
| | **Potassium** | 1,097 | 33 | 33 |
| | **Thiamin** | 3.60 | .11 | .09 |
| | **Riboflavin** | 2.20 | .06 | .06 |
| | **Niacin** | 30.0 | .9 | .9 |
| | **Ascorbic Acid** | Trace | Trace | Trace |
| unit | **Vitamin A** | Trace | Trace | Trace |

*Made with vegetable shortening.

# ENRICHED*
# WHITE BREAD
## (SOFT CRUMB TYPE)

| grams | **Weight:** | 454 | 25 | 22 |
|---|---|---|---|---|

| | | loaf, 1 lb. | slice (18 per loaf) | slice (18 per loaf) toasted |
|---|---|---|---|---|
| % | **Water** | 36 | 36 | 25 |
| | **Calories** | 1,225 | 70 | 70 |
| grams | **Protein** | 39 | 2 | 2 |
| | **Fat** | 15 | 1 | 1 |
| | **Carbohydrate** | 229 | 13 | 13 |
| | **Calcium** | 381 | 21 | 21 |
| milligrams | **Phosphorus** | 440 | 24 | 24 |
| | **Iron** | 11.3 | .6 | .6 |
| | **Sodium** | 2,300 | 127 | 127 |
| | **Potassium** | 476 | 26 | 26 |
| | **Thiamin** | 1.80 | .10 | .08 |
| | **Riboflavin** | 1.10 | .06 | .06 |
| | **Niacin** | 15.0 | .8 | .8 |
| | **Ascorbic Acid** | Trace | Trace | Trace |
| unit | **Vitamin A** | Trace | Trace | Trace |

*Made with vegetable shortening.

362

# ENRICHED*
# WHITE BREAD
## (SOFT CRUMB TYPE, TOASTED)

| | grams | Weight: | 24 | 21 |
|---|---|---|---|---|

| | | loaf, 1½ lb. slice (24 per loaf) | loaf 1½ lb. slice (28 per loaf) |
|---|---|---|---|
| % | Water | 25 | 25 |
| | Calories | 75 | 65 |
| grams | Protein | 2 | 2 |
| | Fat | 1 | 1 |
| | Carbohydrate | 14 | 12 |
| | Calcium | 24 | 20 |
| milligrams | Phosphorus | 27 | 23 |
| | Iron | .7 | .6 |
| | Sodium | 142 | N.A. |
| | Potassium | 29 | 25 |
| | Thiamin | .09 | .08 |
| | Riboflavin | .07 | .06 |
| | Niacin | .9 | .8 |
| | Ascorbic Acid | Trace | Trace |
| unit | Vitamin A | Trace | Trace |

*Made with vegetable shortening.

363

# ENRICHED*
# WHITE BREAD
# (SOFT CRUMB TYPE)

| grams | Weight: | 30 | 45 |
|---|---|---|---|
| | | cubes | crumbs |

| % | Water | 36 | 36 |
|---|---|---|---|
| grams { | Calories | 80 | 120 |
| | Protein | 3 | 4 |
| | Fat | 1 | 1 |
| | Carbohydrate | 15 | 23 |
| | Calcium | 25 | 38 |
| milligrams { | Phosphorus | 29 | 44 |
| | Iron | .8 | 1.1 |
| | Sodium | 152 | 228 |
| | Potassium | 32 | 47 |
| | Thiamin | .12 | .18 |
| | Riboflavin | .07 | .11 |
| | Niacin | 1.0 | 1.5 |
| | Ascorbic Acid | Trace | Trace |
| unit | Vitamin A | Trace | Trace |

*Made with vegetable shortening.

# WHOLE WHEAT BREAD*
## (FIRM CRUMB TYPE)

| grams | **Weight:** | 454 | 25 | 21 |
|---|---|---|---|---|

| | | loaf, 1 lb. | loaf, 1 lb. slice (18 per loaf) | loaf, 1 lb. slice (18 per loaf) toasted |
|---|---|---|---|---|
| % | **Water** | 36 | 36 | 24 |
| | **Calories** | 1,100 | 60 | 60 |
| grams { | **Protein** | 48 | 3 | 3 |
| | **Fat** | 14 | 1 | 1 |
| | **Carbohydrate** | 216 | 12 | 12 |
| | **Calcium** | 449 | 25 | 25 |
| milligrams { | **Phosphorus** | 1,034 | 57 | 57 |
| | **Iron** | 13.6 | .8 | .8 |
| | **Sodium** | 2,390 | 132 | 132 |
| | **Potassium** | 1,238 | 68 | 68 |
| | **Thiamin** | 1.17 | .06 | .05 |
| | **Riboflavin** | .54 | .03 | .03 |
| | **Niacin** | 12.7 | .7 | .7 |
| | **Ascorbic Acid** | Trace | Trace | Trace |
| unit | **Vitamin A** | Trace | Trace | Trace |

*Made with vegetable shortening.

365

# WHOLE WHEAT BREAD*
## (SOFT CRUMB TYPE)

| grams | **Weight:** | 454 | 28 | 24 |
|---|---|---|---|---|

| | | loaf, 1 lb. | loaf, 1 lb. slice (16 per loaf) | loaf, 1 lb. slice (16 per loaf) toasted |
|---|---|---|---|---|
| % | **Water** | 36 | 36 | 24 |
| | **Calories** | 1,095 | 65 | 65 |
| grams | **Protein** | 41 | 3 | 3 |
| | **Fat** | 12 | 1 | 1 |
| | **Carbohydrate** | 224 | 14 | 14 |
| | **Calcium** | 381 | 24 | 24 |
| milligrams | **Phosphorus** | 1,152 | 71 | 71 |
| | **Iron** | 13.6 | .8 | .8 |
| | **Sodium** | 2,404 | 148 | 148 |
| | **Potassium** | 1,161 | 72 | 72 |
| | **Thiamin** | 1.37 | .09 | .07 |
| | **Riboflavin** | .45 | .03 | .03 |
| | **Niacin** | 12.7 | .8 | .8 |
| | **Ascorbic Acid** | Trace | Trace | Trace |
| unit | **Vitamin A** | Trace | Trace | Trace |

*Made with vegetable shortening.

# PASTA

CHOW MEIN NOODLES
EGG NOODLES
MACARONI
MACARONI AND CHEESE
SPAGHETTI
SPAGHETTI IN TOMATO SAUCE WITH CHEESE
SPAGHETTI IN TOMATO SAUCE WITH
MEATBALLS

# CHOW MEIN NOODLES

grams    **Weight:**    45

canned

| % | Water | 1 |
|---|---|---|
| | **Calories** | 220 |
| grams | **Protein** | 6 |
| | **Fat** | 11 |
| | **Carbohydrate** | 26 |
| | **Calcium** | N.A. |
| | **Phosphorus** | N.A. |
| | **Iron** | N.A. |
| milligrams | **Sodium** | N.A. |
| | **Potassium** | N.A. |
| | **Thiamin** | N.A. |
| | **Riboflavin** | N.A. |
| | **Niacin** | N.A. |
| | **Ascorbic Acid** | N.A. |
| unit | **Vitamin A** | N.A. |

# EGG NOODLES*

| grams | **Weight:** | 160 |
|-------|-------------|-----|

cooked

| | | |
|---|---|---|
| % | **Water** | 71 |
| | **Calories** | 200 |
| grams | **Protein** | 7 |
| | **Fat** | 2 |
| | **Carbohydrate** | 37 |
| | **Calcium** | 16 |
| milligrams | **Phosphorus** | 94 |
| | **Iron** | 1.4 |
| | **Sodium** | N.A. |
| | **Potassium** | 70 |
| | **Thiamin** | .22 |
| | **Riboflavin** | .13 |
| | **Niacin** | 1.9 |
| | **Ascorbic Acid** | 0 |
| unit | **Vitamin A** | 110 |

*Enriched

# MACARONI*

| grams | Weight: | 130 | 105 | 140 |
|---|---|---|---|---|

 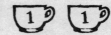 

| | | firm stage: hot | tender stage: cold | tender stage: hot |
|---|---|---|---|---|
| % | Water | 64 | 73 | 73 |
| | Calories | 190 | 115 | 155 |
| grams | Protein | 7 | 4 | 5 |
| | Fat | 1 | Trace | 1 |
| | Carbohydrate | 39 | 24 | 32 |
| | Calcium | 14 | 8 | 11 |
| milligrams | Phosphorus | 85 | 53 | 70 |
| | Iron | 1.4 | .9 | 1.3 |
| | Sodium | **1 | **1 | **1 |
| | Potassium | 103 | 64 | 85 |
| | Thiamin | .23 | .15 | .20 |
| | Riboflavin | .13 | .08 | .11 |
| | Niacin | 1.8 | 1.2 | 1.5 |
| | Ascorbic Acid | 0 | 0 | 0 |
| unit | Vitamin A | 0 | 0 | 0 |

*Enriched
**Value applies to product cooked in unsalted water.

# MACARONI* & CHEESE

| | grams | Weight: | 240 | 200 |
|---|---|---|---|---|

| | | | canned** | from home recipe (served hot)*** |
|---|---|---|---|---|
| % | Water | | 80 | 58 |
| | Calories | | 230 | 430 |
| grams { | Protein | | 9 | 17 |
| | Fat | | 10 | 22 |
| | Carbohydrate | | 26 | 40 |
| | Calcium | | 199 | 362 |
| milligrams { | Phosphorus | | 182 | 322 |
| | Iron | | 1.0 | 1.8 |
| | Sodium | | 730 | 1,086 |
| | Potassium | | 139 | 240 |
| | Thiamin | | .12 | .20 |
| | Riboflavin | | .24 | .40 |
| | Niacin | | 1.0 | 1.8 |
| | Ascorbic Acid | | Trace | Trace |
| unit | Vitamin A | | 260 | 860 |

*Enriched.
**Made with corn oil.
***Made with regular margarine.

# SPAGHETTI*

| | grams | Weight: | 130 | 140 |
|---|---|---|---|---|

| | | firm stage,**<br>served hot | tender stage,<br>served hot |
|---|---|---|---|

| % | **Water** | 64 | 73 |
|---|---|---|---|
| | **Calories** | 190 | 155 |
| grams | **Protein** | 7 | 5 |
| | **Fat** | 1 | 1 |
| | **Carbohydrate** | 39 | 32 |
| | **Calcium** | 14 | 11 |
| milligrams | **Phosphorus** | 85 | 70 |
| | **Iron** | 1.4 | 1.3 |
| | **Sodium** | ***1 | ***1 |
| | **Potassium** | 103 | 85 |
| | **Thiamin** | .23 | .20 |
| | **Riboflavin** | .13 | .11 |
| | **Niacin** | 1.8 | 1.5 |
| | **Ascorbic Acid** | 0 | 0 |
| unit | **Vitamin A** | 0 | 0 |

*Enriched.
**''Aldente''
***Value applies to product cooked in unsalted water.

# SPAGHETTI*
## IN TOMATO SAUCE WITH CHEESE

| | grams | **Weight:** | 250 | 250 |
|---|---|---|---|---|

| | | | home recipe | canned |
|---|---|---|---|---|
| % | | **Water** | 77 | 80 |
| | | **Calories** | 260 | 190 |
| | grams | **Protein** | 9 | 6 |
| | | **Fat** | 9 | 2 |
| | | **Carbohydrate** | 37 | 39 |
| | | **Calcium** | 80 | 40 |
| | | **Phosphorus** | 135 | 88 |
| | | **Iron** | 2.3 | 2.8 |
| | milligrams | **Sodium** | 955 | 955 |
| | | **Potassium** | 408 | 303 |
| | | **Thiamin** | .25 | .35 |
| | | **Riboflavin** | .18 | .28 |
| | | **Niacin** | 2.3 | 4.5 |
| | | **Ascorbic Acid** | 13 | 10 |
| unit | | **Vitamin A** | 1,080 | 930 |

*Enriched.

# SPAGHETTI*
## IN TOMATO SAUCE WITH MEATBALLS

| grams | **Weight:** | 248 | 250 |
|---|---|---|---|

| | | home recipe | canned |
|---|---|---|---|
| % | **Water** | 70 | 78 |
| | **Calories** | 330 | 260 |
| | **Protein** | 19 | 12 |
| grams | **Fat** | 12 | 10 |
| | **Carbohydrate** | 39 | 29 |
| | **Calcium** | 124 | 53 |
| | **Phosphorus** | 236 | 113 |
| | **Iron** | 3.7 | 3.3 |
| milligrams | **Sodium** | 1,009 | 1,220 |
| | **Potassium** | 665 | 245 |
| | **Thiamin** | .25 | .15 |
| | **Riboflavin** | .30 | .18 |
| | **Niacin** | 4.0 | 2.3 |
| | **Ascorbic Acid** | 22 | 5 |
| unit | **Vitamin A** | 1,590 | 1,000 |

*Enriched.

374

# PIES AND PASTRIES

APPLE PIE
BANANA CREAM PIE
BLUEBERRY PIE
CHERRY PIE
CUSTARD PIE
DANISH PASTRY
LEMON MERINGUE PIE
MINCE PIE
PEACH PIE
PECAN PIE
PIE CRUST
PUMPKIN PIE
TOASTER PASTRIES

# APPLE PIE*

| | grams | Weight: | 945 | 158 |
|---|---|---|---|---|

| | whole | sector** |
|---|---|---|

| % | Water | 48 | 47.6 |
|---|---|---|---|
| grams { | Calories | 2,420 | 404 |
| | Protein | 21 | 3.5 |
| | Fat | 105 | 17.5 |
| | Carbohydrate | 360 | 60.2 |
| milligrams { | Calcium | 76 | 13 |
| | Phosphorus | 208 | 35 |
| | Iron | 6.6 | .5 |
| | Sodium | 2,844 | 476 |
| | Potassium | 756 | 126 |
| | Thiamin | 1.06 | .03 |
| | Riboflavin | .79 | .03 |
| | Niacin | 9.3 | .6 |
| | Ascorbic Acid | 9 | 2 |
| unit | Vitamin A | 280 | 50 |

*Pie crust made with enriched flour, vegetable shortening — 9″ diameter.
**Sector is 1/6 of pie.

# BANANA CREAM PIE*

| grams | Weight: | 910 | 130 |
|---|---|---|---|

| | whole | sector** |
|---|---|---|

| | | whole | sector** |
|---|---|---|---|
| % | Water | 54 | 54 |
| | Calories | 2,010 | 285 |
| grams | Protein | 41 | 6 |
| | Fat | 85 | 12 |
| | Carbohydrate | 279 | 40 |
| | Calcium | 601 | 86 |
| milligrams | Phosphorus | 746 | 107 |
| | Iron | 7.3 | 1.0 |
| | Sodium | 840 | 120 |
| | Potassium | 1,847 | 264 |
| | Thiamin | .77 | .11 |
| | Riboflavin | 1.51 | .22 |
| | Niacin | 7.0 | 1.0 |
| | Ascorbic Acid | 9 | 1 |
| unit | Vitamin A | 2,280 | 330 |

*Pie crust made with enriched flour, vegetable shortening — 9" diameter.
**Sector is 1/7 of pie.

# BLUEBERRY PIE*

| | grams | Weight: | 945 | 158 |
|---|---|---|---|---|

| | | whole | sector** |
|---|---|---|---|

| | | | |
|---|---|---|---|
| % | **Water** | 51 | 51.0 |
| grams | **Calories** | 2,285 | 382 |
| | **Protein** | 23 | 3.8 |
| | **Fat** | 102 | 17.1 |
| | **Carbohydrate** | 330 | 55.1 |
| | **Calcium** | 104 | 17 |
| milligrams | **Phosphorus** | 217 | 36 |
| | **Iron** | 9.5 | .9 |
| | **Sodium** | 2,533 | 423 |
| | **Potassium** | 614 | 103 |
| | **Thiamin** | 1.03 | .03 |
| | **Riboflavin** | .80 | .03 |
| | **Niacin** | 10.0 | .5 |
| | **Ascorbic Acid** | 28 | 5 |
| unit | **Vitamin A** | 280 | 50 |

*Pie crust made with enriched flour, vegetable shortening — 9″ diameter.
**Sector is 1/6 of pie.

# CHERRY PIE*

| | | |
|---|---|---|
| grams | **Weight:** | 945 | 158 |

whole    sector**

| | whole | sector** |
|---|---|---|
| % | **Water** | 47 | 46.6 |
| | **Calories** | 2,465 | 412 |
| grams | **Protein** | 25 | 4.1 |
| | **Fat** | 107 | 17.9 |
| | **Carbohydrate** | 363 | 60.7 |
| | **Calcium** | 132 | 22 |
| milligrams | **Phosphorus** | 236 | 40 |
| | **Iron** | 6.6 | .5 |
| | **Sodium** | 2,873 | 480 |
| | **Potassium** | 992 | 166 |
| | **Thiamin** | 1.09 | .03 |
| | **Riboflavin** | .84 | .03 |
| | **Niacin** | 9.8 | .8 |
| | **Ascorbic Acid** | Trace | Trace |
| unit | **Vitamin A** | 4,160 | 690 |

*Pie crust made with enriched flour, vegetable shortening — 9″ diameter.
**Section is 1/6 of pie.

# CUSTARD PIE*

| | grams | Weight: | 910 | 130 |
|---|---|---|---|---|

| | whole | sector** |
|---|---|---|

| | | whole | sector** |
|---|---|---|---|
| % | **Water** | 58 | 58 |
| | **Calories** | 1,985 | 285 |
| grams | **Protein** | 56 | 8 |
| | **Fat** | 101 | 14 |
| | **Carbohydrate** | 213 | 30 |
| | **Calcium** | 874 | 125 |
| milligrams | **Phosphorus** | 1,028 | 147 |
| | **Iron** | 8.2 | 1.2 |
| | **Sodium** | 2,612 | 436 |
| | **Potassium** | 1,247 | 178 |
| | **Thiamin** | .79 | .11 |
| | **Riboflavin** | 1.92 | .27 |
| | **Niacin** | 5.6 | .8 |
| | **Ascorbic Acid** | 0 | 0 |
| unit | **Vitamin A** | 2,090 | 300 |

*Pie crust made with enriched flour, vegetable shortening — 9″ diameter.
**Sector is 1/7 of pie.

# WITHOUT FRUIT OR NUTS*
# DANISH PASTRY
## (ENRICHED FLOUR)

| grams | Weight: | 340 | 65 | 28 |
|---|---|---|---|---|
| | | **1** packaged ring, 12 oz. | **1** round piece about 4¼" diam. by 1" | OZ **1** ounce |

| | | | | |
|---|---|---|---|---|
| % | **Water** | 22 | 22 | 22 |
| | **Calories** | 1,435 | 275 | 120 |
| grams | **Protein** | 25 | 5 | 2 |
| | **Fat** | 80 | 15 | 7 |
| | **Carbohydrate** | 155 | 30 | 13 |
| | **Calcium** | 170 | 33 | 14 |
| milligrams | **Phosphorus** | 371 | 71 | 31 |
| | **Iron** | 6.1 | 1.2 | .5 |
| | **Sodium** | 1,244 | 238 | 104 |
| | **Potassium** | 381 | 73 | 32 |
| | **Thiamin** | .97 | .18 | .08 |
| | **Riboflavin** | 1.01 | .19 | .08 |
| | **Niacin** | 8.6 | 1.7 | .7 |
| | **Ascorbic Acid** | Trace | Trace | Trace |
| unit | **Vitamin A** | 1,050 | 200 | 90 |

*Contains vegetable shortening and butter.

# LEMON
# MERINGUE PIE*

| | **Weight:** | 840 | 140 |
|---|---|---|---|
| grams | | | |

| | | whole | sector** |
|---|---|---|---|

| | | | |
|---|---|---|---|
| % | **Water** | 47 | 47.4 |
| grams { | **Calories** | 2,140 | 357 |
| | **Protein** | 31 | 5.2 |
| | **Fat** | 86 | 14.3 |
| | **Carbohydrate** | 317 | 52.8 |
| | **Calcium** | 118 | 20 |
| milligrams { | **Phosphorus** | 412 | 69 |
| | **Iron** | 6.7 | .7 |
| | **Sodium** | 2,369 | 395 |
| | **Potassium** | 420 | 70 |
| | **Thiamin** | .61 | .04 |
| | **Riboflavin** | .84 | .11 |
| | **Niacin** | 5.2 | .3 |
| | **Ascorbic Acid** | 25 | 4 |
| unit | **Vitamin A** | 1,430 | 240 |

*Pie crust made with enriched flour, vegetable shortening — 9″ diameter.
**Sector is 1/6 of pie.

# MINCE PIE*

| | | whole | sector** |
|---|---|---|---|
| grams | **Weight:** | 945 | 158 |

| | | whole | sector** |
|---|---|---|---|
| % | **Water** | 43 | 43.0 |
| | **Calories** | 2,560 | 428 |
| grams | **Protein** | 24 | 4.0 |
| | **Fat** | 109 | 18.2 |
| | **Carbohydrate** | 389 | 65.1 |
| | **Calcium** | 265 | 44 |
| milligrams | **Phosphorus** | 359 | 60 |
| | **Iron** | 13.3 | 1.6 |
| | **Sodium** | 4,234 | 708 |
| | **Potassium** | 1,682 | 281 |
| | **Thiamin** | .96 | .11 |
| | **Riboflavin** | .86 | .06 |
| | **Niacin** | 9.8 | .6 |
| | **Ascorbic Acid** | 9 | 2 |
| unit | **Vitamin A** | 20 | Trace |

*Pie crust made with enriched flour, vegetable shortening — 9″ diameter.
**Sector is 1/6 of pie.

# PEACH PIE*

| | grams | Weight: | 945 | 158 |
|---|---|---|---|---|

| | | whole | sector** |
|---|---|---|---|

| | | | |
|---|---|---|---|
| % | **Water** | 48 | 47.5 |
| | **Calories** | 2,410 | 403 |
| grams { | **Protein** | 24 | 4.0 |
| | **Fat** | 101 | 16.9 |
| | **Carbohydrate** | 361 | 60.4 |
| | **Calcium** | 95 | 16 |
| milligrams { | **Phosphorus** | 274 | 46 |
| | **Iron** | 8.5 | .8 |
| | **Sodium** | 2,533 | 423 |
| | **Potassium** | 1,408 | 235 |
| | **Thiamin** | 1.04 | .03 |
| | **Riboflavin** | .97 | .06 |
| | **Niacin** | 14.0 | 1.1 |
| | **Ascorbic Acid** | 28 | 5 |
| unit | **Vitamin A** | 6,900 | 1,150 |

*Pie crust made with enriched flour, vegetable shortening — 9″ diameter.
**Sector is 1/6 of pie.

# PECAN PIE*

|  | **Weight:** | 825 | 138 |
|---|---|---|---|
| grams | | | |

|  | whole | sector** |
|---|---|---|

| | | whole | sector** |
|---|---|---|---|
| % | **Water** | 20 | 19.5 |
| | **Calories** | 3,450 | 577 |
| grams | **Protein** | 42 | 7.0 |
| | **Fat** | 189 | 31.6 |
| | **Carbohydrate** | 423 | 70.8 |
| | **Calcium** | 388 | 65 |
| milligrams | **Phosphorus** | 850 | 142 |
| | **Iron** | 25.6 | 3.9 |
| | **Sodium** | 1,823 | 305 |
| | **Potassium** | 1,015 | 170 |
| | **Thiamin** | 1.80 | .22 |
| | **Riboflavin** | .95 | .10 |
| | **Niacin** | 6.9 | .4 |
| | **Ascorbic Acid** | Trace | Trace |
| unit | **Vitamin A** | 1,320 | 220 |

*Pie crust made with enriched flour, vegetable shortening — 9" diameter.
**Sector is 1/6 of pie.

# PIE CRUST

| | Weight: | 180 | 320 |
|---|---|---|---|
| grams | | **1** | **1** |
| | | home recipe* 9″ diam. | 2 crust pie 9″ diam. each** |
| % | **Water** | 15 | 19 |
| | **Calories** | 900 | 1,485 |
| grams | **Protein** | 11 | 20 |
| | **Fat** | 60 | 93 |
| | **Carbohydrate** | 79 | 141 |
| | **Calcium** | 25 | 131 |
| milligrams | **Phosphorus** | 90 | 272 |
| | **Iron** | 3.1 | 6.1 |
| | **Sodium** | 1,102 | 2,602 |
| | **Potassium** | 89 | 179 |
| | **Thiamin** | .47 | 1.07 |
| | **Riboflavin** | .40 | .79 |
| | **Niacin** | 5.0 | 9.9 |
| | **Ascorbic Acid** | 0 | 0 |
| unit | **Vitamin A** | 0 | 0 |

*Made with enriched flour and vegetable shortening, baked.
**10 oz. pkg. prepared and baked. Mix with enriched flour and vegetable shortening.

# PUMPKIN PIE*

| | Weight: | 910 | 152 |
|---|---|---|---|
| grams | | | |

| | | whole | sector** |
|---|---|---|---|

| | | | |
|---|---|---|---|
| % | **Water** | 59 | 59.2 |
| | **Calories** | 1,920 | 321 |
| grams | **Protein** | 36 | 6.1 |
| | **Fat** | 102 | 17.0 |
| | **Carbohydrate** | 223 | 37.2 |
| | **Calcium** | 464 | 78 |
| milligrams | **Phosphorus** | 628 | 105 |
| | **Iron** | 7.3 | .8 |
| | **Sodium** | 1,947 | 325 |
| | **Potassium** | 1,456 | 243 |
| | **Thiamin** | .78 | .05 |
| | **Riboflavin** | 1.27 | .15 |
| | **Niacin** | 7.0 | .8 |
| | **Ascorbic Acid** | Trace | Trace |
| unit | **Vitamin A** | 22,480 | 3,750 |

*Pie crust made with enriched flour, vegetable shortening — 9" diameter.
**Sector is 1/6 of pie:

# TOASTER PASTRIES

grams **Weight:** .50

| | |
|---|---|
| % | **Water** 12 |
| grams | **Calories** 200 |
| | **Protein** 3 |
| | **Fat** 6 |
| | **Carbohydrate** 36 |
| | **Calcium** *54 |
| milligrams | **Phosphorus** *67 |
| | **Iron** 1.9 |
| | **Sodium** N.A. |
| | **Potassium** *74 |
| | **Thiamin** .16 |
| | **Riboflavin** .17 |
| | **Niacin** 2.1 |
| | **Ascorbic Acid** * |
| unit | **Vitamin A** 500 |

*Value varies with the brand. Consult the label.

# CAKES AND ICINGS

ANGELFOOD CAKE
BOSTON CREAM PIE
COFFEE CAKE
CUPCAKES, WHITE AND DEVIL'S FOOD
DEVIL'S FOOD CAKE
FRUITCAKE
GINGERBREAD CAKE
ICING, BOILED
ICING, UNCOOKED
MARBLE CAKE
PLAIN SHEETCAKE
POUND CAKE
SPONGE CAKE
WHITE CAKE
YELLOW CAKE

# ANGELFOOD CAKE

| | grams | **Weight:** | 716 | 60 |
|---|---|---|---|---|

| | | whole cake 9¾" diam. tube | piece ¹⁄₁₂ of cake |
|---|---|---|---|

| % | **Water** | 31.5 | 31.5 |
|---|---|---|---|
| grams | **Calories** | 1,926 | 161 |
| | **Protein** | 50.8 | 4.3 |
| | **Fat** | 1.4 | .1 |
| | **Carbohydrate** | 431.0 | 36.1 |
| | **Calcium** | 64 | 5 |
| milligrams | **Phosphorus** | 158 | 13 |
| | **Iron** | 1.4 | .1 |
| | **Sodium** | 2,026 | 170 |
| | **Potassium** | 630 | 53 |
| | **Thiamin** | .07 | .01 |
| | **Riboflavin** | 1.00 | .08 |
| | **Niacin** | 1.4 | .1 |
| | **Ascorbic Acid** | 0 | 0 |
| unit | **Vitamin A** | 0 | 0 |

390

# BOSTON CREAM PIE
## WITH CUSTARD FILLING

| | grams | Weight: | 825 | 69 |

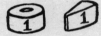

| | whole cake 8″ diam. | piece 1/12 of cake |

| | | | whole cake 8″ diam. | piece 1/12 of cake |
|---|---|---|---|---|
| % | **Water** | | 35 | 35 |
| | **Calories** | | 2,490 | 210 |
| grams | **Protein** | | 41 | 3 |
| | **Fat** | | 78 | 6 |
| | **Carbohydrate** | | 412 | 34 |
| | **Calcium** | | 553 | 46 |
| milligrams | **Phosphorus** | | 833 | 70 |
| | **Iron** | | 8.2 | .7 |
| | **Sodium** | | 1,535 | 128 |
| | **Potassium** | | *734 | *61 |
| | **Thiamin** | | 1.04 | .09 |
| | **Riboflavin** | | 1.27 | .11 |
| | **Niacin** | | 9.6 | .8 |
| | **Ascorbic Acid** | | 2 | Trace |
| unit | **Vitamin A** | | 1,730 | 140 |

If butter or margarine used for cake portion, vitamin A would be higher.
*Applies to product made with a sodium-sulfate type baking powder.

# COFFEE CAKE

| | | |
|---|---|---|
| grams | **Weight:** | 430 | 72 |

| | whole cake<br>7¾" × 5⅝"<br>× 1¼" | piece,<br>¹⁄₁₂ of cake |
|---|---|---|
| % **Water** | 30 | 30 |
| **Calories** | 1,385 | 230 |
| **Protein** | 27 | 5 |
| **Fat** | 41 | 7 |
| **Carbohydrate** | 225 | 38 |
| **Calcium** | 262 | 44 |
| **Phosphorus** | 748 | 125 |
| **Iron** | 6.9 | 1.2 |
| **Sodium** | 1,853 | 310 |
| **Potassium** | 469 | 78 |
| **Thiamin** | .82 | .14 |
| **Riboflavin** | .91 | .15 |
| **Niacin** | 7.7 | 1.3 |
| **Ascorbic Acid** | 1 | Trace |
| **Vitamin A** | 690 | 120 |

Made from enriched flour mix and vegetable shortening.

392

# CUPCAKES
## WHITE & DEVIL'S FOOD

| grams | **Weight:** | 25 | 36 | 35 |
|---|---|---|---|---|
| | | white, without icing, 2½″ diam. | white, with chocolate icing, 2½″ diam. | devil's food with choc. icing, 2½″ diam. |
| % | **Water** | 26 | 22 | 24 |
| grams | **Calories** | 90 | 130 | 120 |
| | **Protein** | 1 | 2 | 2 |
| | **Fat** | 3 | 5 | 4 |
| | **Carbohydrate** | 14 | 21 | 20 |
| | **Calcium** | 40 | 47 | 21 |
| milligrams | **Phosphorus** | 59 | 71 | 37 |
| | **Iron** | .3 | .4 | .5 |
| | **Sodium** | 113 | 121 | 92 |
| | **Potassium** | 21 | 42 | 46 |
| | **Thiamin** | .05 | .05 | .03 |
| | **Riboflavin** | .05 | .06 | .05 |
| | **Niacin** | .4 | .4 | .3 |
| | **Ascorbic Acid** | Trace | Trace | Trace |
| unit | **Vitamin A** | 40 | 60 | 50 |

Made from mix using enriched flour and vegetable shortening. Icing made with butter.

# DEVIL'S FOOD CAKE
## (WITH CHOCOLATE FROSTING)

| | grams | **Weight:** | 1,107 | 69 |
|---|---|---|---|---|

| | | whole, 2 layer cake (8 or 9″ diam.) | piece 1/16 of cake |
|---|---|---|---|
| % | **Water** | 24 | 24 |
| | **Calories** | 3,755 | 235 |
| grams | **Protein** | 49 | 3 |
| | **Fat** | 136 | 8 |
| | **Carbohydrate** | 645 | 40 |
| | **Calcium** | 653 | 41 |
| milligrams | **Phosphorus** | 1,162 | 72 |
| | **Iron** | 16.6 | 1.0 |
| | **Sodium** | 2,900 | 181 |
| | **Potassium** | 1,439 | 90 |
| | **Thiamin** | 1.06 | .07 |
| | **Riboflavin** | 1.65 | .10 |
| | **Niacin** | 10.1 | .6 |
| | **Ascorbic Acid** | 1 | Trace |
| unit | **Vitamin A** | 1,660 | 100 |

Made from mix with enriched flour and vegetable shortening. Icing made with butter.

# FRUITCAKE
## DARK

| | grams Weight: | 454 | 15 |
|---|---|---|---|

| | | loaf, 1 lb.<br>7½" x 2"<br>x 1½" | slice,<br>1/30<br>of loaf |
|---|---|---|---|
| % | Water | 18 | 18 |
| | Calories | 1,720 | 55 |
| grams | Protein | 22 | 1 |
| | Fat | 64 | 2 |
| | Carbohydrate | 271 | 9 |
| | Calcium | 327 | 11 |
| milligrams | Phosphorus | 513 | 17 |
| | Iron | 11.8 | .4 |
| | Sodium | 717 | 24 |
| | Potassium | 2,250 | 74 |
| | Thiamin | .72 | .02 |
| | Riboflavin | .73 | .02 |
| | Niacin | 4.9 | .2 |
| | Ascorbic Acid | 2 | Trace |
| unit | Vitamin A | 540 | 20 |

From home recipe using enriched flour and vegetable shortening.

# GINGERBREAD CAKE

| | grams | **Weight:** | 570 | 63 |
|---|---|---|---|---|

| | | whole cake (8″ square) | piece, 1/9 of cake |
|---|---|---|---|

| | | whole cake (8″ square) | piece, 1/9 of cake |
|---|---|---|---|
| % | Water | 37 | 37 |
| | Calories | 1,575 | 175 |
| grams | Protein | 18 | 2 |
| | Fat | 39 | 4 |
| | Carbohydrate | 291 | 32 |
| | Calcium | 513 | 57 |
| | Phosphorus | 570 | 63 |
| | Iron | 8.6 | .9 |
| milligrams | Sodium | 1,733 | 192 |
| | Potassium | 1,562 | 173 |
| | Thiamin | .84 | .09 |
| | Riboflavin | 1.00 | .11 |
| | Niacin | 7.4 | .8 |
| | Ascorbic Acid | Trace | Trace |
| unit | Vitamin A | Trace | Trace |

Made from mix with enriched flour and vegetable shortening.

# CAKE ICING
## BOILED

| grams | **Weight:** | 94 | 166 |
|---|---|---|---|

| | | white, plain | white, with coconut |
|---|---|---|---|
| % | **Water** | 18 | 15 |
| | **Calories** | 295 | 605 |
| grams | **Protein** | 1 | 3 |
| | **Fat** | 0 | 13 |
| | **Carbohydrate** | 75 | 124 |
| | **Calcium** | 2 | 10 |
| milligrams | **Phosphorus** | 2 | 50 |
| | **Iron** | Trace | .8 |
| | **Sodium** | 134 | 196 |
| | **Potassium** | 17 | 277 |
| | **Thiamin** | Trace | .02 |
| | **Riboflavin** | .03 | .07 |
| | **Niacin** | Trace | .3 |
| | **Ascorbic Acid** | 0 | 0 |
| unit | **Vitamin A** | 0 | 0 |

# CAKE ICING
## UNCOOKED

| | grams | Weight: | 275 | 245 | 319 |
|---|---|---|---|---|---|

| | | chocolate made with milk & butter | creamy fudge from mix & water | white |
|---|---|---|---|---|
| % | Water | 14 | 15 | 11 |
| | Calories | 1,035 | 830 | 1,200 |
| grams | Protein | 9 | 7 | 2 |
| | Fat | 38 | 16 | 21 |
| | Carbohydrate | 185 | 183 | 260 |
| | Calcium | 165 | 96 | 48 |
| milligrams | Phosphorus | 305 | 218 | 38 |
| | Iron | 3.3 | 2.7 | Trace |
| | Sodium | 168 | 568 | 156 |
| | Potassium | 536 | 238 | 57 |
| | Thiamin | .06 | .05 | Trace |
| | Riboflavin | .28 | .20 | .06 |
| | Niacin | .6 | .7 | Trace |
| | Ascorbic Acid | 1 | Trace | Trace |
| unit | Vitamin A | 580 | Trace | 860 |

# MARBLE CAKE
## WITH BOILED WHITE ICING

| | Weight: | 1,045 | 87 |
|---|---|---|---|
| grams | | | |

| | | whole cake, 2 layer, 9″ diam. | piece, 1/12 of cake |
|---|---|---|---|
| % | Water | 23.6 | 23.6 |
| | Calories | 3,459 | 288 |
| grams | Protein | 46 | 3.8 |
| | Fat | 90.9 | 7.6 |
| | Carbohydrate | 647.9 | 53.9 |
| | Calcium | 815 | 68 |
| milligrams | Phosphorus | 1,787 | 149 |
| | Iron | 8.4 | .7 |
| | Sodium | 2,707 | 225 |
| | Potassium | 1,275 | 106 |
| | Thiamin | .21 | .02 |
| | Riboflavin | .84 | .07 |
| | Niacin | 2.1 | .2 |
| | Ascorbic Acid | 1 | Trace |
| unit | Vitamin A | 940 | 80 |

399

# PLAIN SHEETCAKE

| | grams | **Weight:** | 777 | 86 |
|---|---|---|---|---|

| | | without icing: whole cake (9″ square) | without icing piece, 1/9 of cake |
|---|---|---|---|
| % | **Water** | 25 | 25 |
| grams | **Calories** | 2,830 | 315 |
| | **Protein** | 35 | 4 |
| | **Fat** | 108 | 12 |
| | **Carbohydrate** | 434 | 48 |
| | **Calcium** | 497 | 55 |
| milligrams | **Phosphorus** | 793 | 88 |
| | **Iron** | 8.5 | .9 |
| | **Sodium** | 2,331 | 258 |
| | **Potassium** | *614 | *68 |
| | **Thiamin** | 1.21 | .13 |
| | **Riboflavin** | 1.40 | .15 |
| | **Niacin** | 10.2 | 1.1 |
| | **Ascorbic Acid** | 2 | Trace |
| unit | **Vitamin A** | 1,320 | 150 |

Made from home recipe using enriched flour and vegetable shortening. Icing made with butter.
*Applies to product made with a sodium aluminum-sulfate type baking powder.

# PLAIN SHEETCAKE

| | grams | | |
|---|---|---|---|
| | **Weight:** | 1,096 | 121 |

| | | with uncooked white icing; whole cake (9″ square) | with uncooked white icing; piece 1/9 of cake |
|---|---|---|---|
| % | **Water** | 21 | 21 |
| grams { | **Calories** | 4,020 | 445 |
| | **Protein** | 37 | 4 |
| | **Fat** | 129 | 14 |
| | **Carbohydrate** | 694 | 77 |
| milligrams { | **Calcium** | 548 | 61 |
| | **Phosphorus** | 822 | 91 |
| | **Iron** | 8.2 | .8 |
| | **Sodium** | 2,488 | 275 |
| | **Potassium** | *669 | *74 |
| | **Thiamin** | 1.22 | .14 |
| | **Riboflavin** | 1.47 | .16 |
| | **Niacin** | 10.2 | 1.1 |
| | **Ascorbic Acid** | 2 | Trace |
| unit | **Vitamin A** | 2,190 | 240 |

Made from home recipe using enriched flour and vegetable shortening. Icing made with butter.
*Applies to product made with a sodium aluminum-sulfate type baking powder.

# POUND CAKE

| grams | Weight: | 514 | 30 |
|---|---|---|---|

| | | loaf;<br>8½ x 3½ x<br>3″ | slice; 3½ x<br>3 x ½″;<br>1/17 of loaf |
|---|---|---|---|
| % | Water | 17.2 | 17.2 |
| | Calories | 2,431 | 142 |
| grams { | Protein | 29.3 | 1.7 |
| | Fat | 151.6 | 8.9 |
| | Carbohydrate | 241.6 | 14.1 |
| | Calcium | 108 | 6 |
| milligrams { | Phosphorus | 406 | 24 |
| | Iron | 4.1 | .2 |
| | Sodium | 565 | 33 |
| | Potassium | 308 | 18 |
| | Thiamin | .15 | .01 |
| | Riboflavin | .46 | .03 |
| | Niacin | 1.0 | .1 |
| | Ascorbic Acid | 0 | 0 |
| unit | Vitamin A | 1,440 | 80 |

Made from home recipe using enriched flour and vegetable shortening.

# SPONGE CAKE

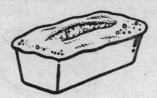

| | grams | Weight: | 790 | 66 |
|---|---|---|---|---|

9¾″ diam. tube     1/12 of cake

| | | | |
|---|---|---|---|
| % | **Water** | 32 | 32 |
| | **Calories** | 2,345 | 195 |
| grams | **Protein** | 60 | 5 |
| | **Fat** | 45 | 4 |
| | **Carbohydrate** | 427 | 36 |
| | **Calcium** | 237 | 20 |
| | **Phosphorus** | 885 | 74 |
| | **Iron** | 9.5 | .8 |
| milligrams | **Sodium** | 1,319. | 110 |
| | **Potassium** | 687 | 57 |
| | **Thiamin** | 1.10 | .09 |
| | **Riboflavin** | 1.64 | .14 |
| | **Niacin** | 7.4 | .6 |
| | **Ascorbic Acid** | Trace | Trace |
| unit | **Vitamin A** | 3,560 | 300 |

Made from home recipe using enriched flour.

# WHITE CAKE
# (WITH CHOCOLATE FROSTING)

| | | whole cake* (8 or 9" diam.) | piece 1/16 of cake |
|---|---|---|---|
| grams | **Weight:** | 1,140 | 71 |

| | | whole cake* (8 or 9" diam.) | piece 1/16 of cake |
|---|---|---|---|
| % | **Water** | 21 | 21 |
| | **Calories** | 4,000 | 250 |
| grams | **Protein** | 44 | 3 |
| | **Fat** | 122 | 8 |
| | **Carbohydrate** | 716 | 45 |
| | **Calcium** | 1,129 | 70 |
| milligrams | **Phosphorus** | 2,041 | 127 |
| | **Iron** | 11.4 | .7 |
| | **Sodium** | 2,588 | 161 |
| | **Potassium** | 1,322 | 82 |
| | **Thiamin** | 1.50 | .09 |
| | **Riboflavin** | 1.77 | .11 |
| | **Niacin** | 12.5 | .8 |
| | **Ascorbic Acid** | 2 | Trace |
| unit | **Vitamin A** | 680 | 40 |

*2 layer with chocolate icing. Cake made from mix with enriched flour and vegetable shortening. Icing made with butter.

# YELLOW CAKE
## (WITH CHOCOLATE FROSTING)

| grams — | **Weight:** | 1,108 | 69 |
|---|---|---|---|

| | | whole cake*<br>(8 or 9"<br>diam) | piece*<br>1/16 of cake |
|---|---|---|---|
| % | **Water** | 26 | 26 |
| | **Calories** | 3,735 | 235 |
| | **Protein** | 45 | 3 |
| | **Fat** | 125 | 8 |
| | **Carbohydrate** | 638 | 40 |
| | **Calcium** | 1,008 | 63 |
| | **Phosphorus** | 2,017 | 126 |
| | **Iron** | 12.2 | .8 |
| | **Sodium** | 2,515 | 157 |
| | **Potassium** | 1,208 | 75 |
| | **Thiamin** | 1.24 | .08 |
| | **Riboflavin** | 1.67 | .10 |
| | **Niacin** | 10.6 | .7 |
| | **Ascorbic Acid** | 2 | Trace |
| unit | **Vitamin A** | 1,550 | 100 |

*grams {* applies to Protein through Carbohydrate; *milligrams {* applies to Calcium through Vitamin A*

*2 layer with chocolate icing. Cake made from mix with enriched flour and vegetable shortening. Icing made with butter.

405

# COOKIES AND DOUGHNUTS

BROWNIES
CHOCOLATE CHIP COOKES
DOUGHNUTS
FIG BARS
GINGERSNAPS
MACAROONS
OATMEAL COOKIES
SANDWICH TYPE COOKIES
VANILLA WAFERS

# BROWNIES
## WITH NUTS

| | grams Weight: | 20 | 20 | 25 |
|---|---|---|---|---|

| | | home prepared from home recipe 1¾″ x 1¾″ x ⅞″ | home prepared from commercial recipe 1¾″ x 1¾″ x ⅞″ | frozen with chocolate icing 1½″ x 1¾″ x ⅞″ |
|---|---|---|---|---|
| % | **Water** | 10 | 11 | 13 |
| grams | **Calories** | 95 | 85 | 105 |
| | **Protein** | 1 | 1 | 1 |
| | **Fat** | 6 | 4 | 5 |
| | **Carbohydrate** | 10 | 13 | 15 |
| | **Calcium** | 8 | 9 | 10 |
| milligrams | **Phosphorus** | 30 | 27 | 31 |
| | **Iron** | .4 | .4 | .4 |
| | **Sodium** | 50 | N.A. | N.A. |
| | **Potassium** | 38 | 34 | 44 |
| | **Thiamin** | .04 | .03 | .03 |
| | **Riboflavin** | .03 | .02 | .03 |
| | **Niacin** | .2 | .2 | .2 |
| | **Ascorbic Acid** | Trace | Trace | Trace |
| unit | **Vitamin A** | 40 | 20 | 50 |

Made wtih enriched flour and vegetable shortening. Icing made with butter.

# CHOCOLATE CHIP
## COOKIES*

| grams | Weight: | 42 | 40 | 48 |
|---|---|---|---|---|
| | | commercial 2¼" diam. | home recipe 2⅓" diam. | commercial chilled dough, 2½" diam. |

| | | | | |
|---|---|---|---|---|
| % | **Water** | 3 | 3 | 5 |
| grams | **Calories** | 200 | 205 | 240 |
| | **Protein** | 2 | 2 | 2 |
| | **Fat** | 9 | 12 | 12 |
| | **Carbohydrate** | 29 | 24 | 31 |
| | **Calcium** | 16 | 14 | 17 |
| milligrams | **Phosphorus** | 48 | 40 | 35 |
| | **Iron** | 1.0 | .8 | .6 |
| | **Sodium** | 168 | 139 | N.A. |
| | **Potassium** | 56 | 47 | 23 |
| | **Thiamin** | .10 | .06 | .10 |
| | **Riboflavin** | .17 | .06 | .08 |
| | **Niacin** | .9 | .5 | .9 |
| | **Ascorbic Acid** | Trace | Trace | 0 |
| unit | **Vitamin A** | 50 | 40 | 30 |

*Made with enriched flour and vegetable shortening.

# DOUGHNUTS

| | Weight: | 25 | 50 |
|---|---|---|---|
grams

| | | cake type, plain, 2½″ diam. 1″ high* | yeast-leavened, glazed, 3¾″ diam., 1¼″ high |
|---|---|---|---|
| % | **Water** | 24 | 26 |
| | **Calories** | 100 | 205 |
| grams | **Protein** | 1 | 1 |
| | **Fat** | 5 | 5 |
| | **Carbohydrate** | 13 | 13 |
| | **Calcium** | 10 | 10 |
| | **Phosphorus** | 48 | 48 |
| | **Iron** | .4 | .4 |
| milligrams | **Sodium** | 125 | N.A. |
| | **Potassium** | 23 | 23 |
| | **Thiamin** | .05 | .05 |
| | **Riboflavin** | .05 | .05 |
| | **Niacin** | .4 | .4 |
| | **Ascorbic Acid** | Trace | Trace |
| unit | **Vitamin A** | 20 | 20 |

*Made with vegetable shortening and enriched flour.

409

# FIG BARS

| | | |
|---|---|---|
| grams | **Weight:** | 56 |

**4**

commercial*
1½" x 1¾"
x ½"

| | | |
|---|---|---|
| % | **Water** | 14 |
| grams | **Calories** | 200 |
| | **Protein** | 2 |
| | **Fat** | 3 |
| | **Carbohydrate** | 42 |
| | **Calcium** | 44 |
| milligrams | **Phosphorus** | 34 |
| | **Iron** | 1.0 |
| | **Sodium** | 141 |
| | **Potassium** | 111 |
| | **Thiamin** | .04 |
| | **Riboflavin** | .14 |
| | **Niacin** | .9 |
| | **Ascorbic Acid** | Trace |
| unit | **Vitamin A** | 60 |

*Made with enriched flour and vegetable shortening.

# GINGERSNAPS*

grams | **Weight:** | 70

2″ diam.,
¼″ thick

| | | |
|---|---|---|
| % | **Water** | 3.1 |
| | **Calories** | 294 |
| grams | **Protein** | 3.9 |
| | **Fat** | 6.2 |
| | **Carbohydrate** | 55.9 |
| | **Calcium** | 51 |
| | **Phosphorus** | 33 |
| | **Iron** | 1.6 |
| | **Sodium** | 400 |
| milligrams | **Potassium** | 323 |
| | **Thiamin** | .03 |
| | **Riboflavin** | .04 |
| | **Niacin** | .3 |
| | **Ascorbic Acid** | Trace |
| unit | **Vitamin A** | 50 |

*Commercial, made with enriched flour and vegetable shortening.

# MACAROONS

grams **Weight:** 38

2¾″ diam.,
¼″ thick*

| | % | |
|---|---|---|
| Water | 4 | |

| | grams | |
|---|---|---|
| Calories | 180 | |
| Protein | 2 | |
| Fat | 9 | |
| Carbohydrate | 25 | |

| | milligrams | |
|---|---|---|
| Calcium | 10 | |
| Phosphorus | 32 | |
| Iron | .3 | |
| Sodium | 13 | |
| Potassium | 176 | |
| Thiamin | .02 | |
| Riboflavin | .06 | |
| Niacin | .2 | |
| Ascorbic Acid | 0 | |
| Vitamin A | 0 | unit |

*Made with enriched flour.

# OATMEAL COOKIES
## WITH RAISINS

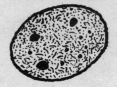

grams    **Weight:**      52

2⅝″ diam.,
¼″ thick*

| | | |
|---|---|---|
| % | **Water** | 3 |
| | **Calories** | 235 |
| | **Protein** | 3 |
| grams | **Fat** | 8 |
| | **Carbohydrate** | 38 |
| | **Calcium** | 11 |
| | **Phosphorus** | 53 |
| | **Iron** | 1.4 |
| | **Sodium** | 84 |
| milligrams | **Potassium** | 192 |
| | **Thiamin** | .15 |
| | **Riboflavin** | .10 |
| | **Niacin** | 1.0 |
| | **Ascorbic Acid** | Trace |
| unit | **Vitamin A** | 30 |

*Commercial, made with enriched flour and vegetable shortening.

# SANDWICH TYPE
## CHOCOLATE OR VANILLA

grams **Weight:** 40

1¾" diam.*

| | | |
|---|---|---|
| % | **Water** | 2 |
| grams | **Calories** | 200 |
| | **Protein** | 2 |
| | **Fat** | 9 |
| | **Carbohydrate** | 28 |
| | **Calcium** | 10 |
| milligrams | **Phosphorus** | 96 |
| | **Iron** | .7 |
| | **Sodium** | 193 |
| | **Potassium** | 15 |
| | **Thiamin** | .06 |
| | **Riboflavin** | .10 |
| | **Niacin** | .7 |
| | **Ascorbic Acid** | 0 |
| unit | **Vitamin A** | 0 |

*Commercial, made with enriched flour and vegetable shortening.

# VANILLA WAFERS

grams **Weight:** 40

1¾″ diam.*

| | | |
|---|---|---|
| % | **Water** | 3 |
| | **Calories** | 185 |
| grams { | **Protein** | 2 |
| | **Fat** | 6 |
| | **Carbohydrate** | 30 |
| | **Calcium** | 16 |
| milligrams { | **Phosphorus** | 25 |
| | **Iron** | .6 |
| | **Sodium** | 101 |
| | **Potassium** | 29 |
| | **Thiamin** | .10 |
| | **Riboflavin** | .09 |
| | **Niacin** | .8 |
| | **Ascorbic Acid** | 0 |
| unit | **Vitamin A** | 50 |

*Commercial, made with enriched flour and vegetable shortening.

# BREAKFAST FOODS

READY-TO-EAT CEREALS
BRAN FLAKES
CORN FLAKES
PUFFED CORN
PUFFED OATS
PUFFED RICE
PUFFED WHEAT
SHREDDED CORN
SHREDDED WHEAT
WHEATGERM, TOASTED
WHEAT FLAKES

COOKED, HOT CEREALS
CORN GRITS
FARINA
OATMEAL OR ROLLED OATS
ROLLED WHEAT
WHOLE WHEAT MEAL

MISC. BREAKFAST FOODS
PANCAKES
WAFFLES

# BRAN FLAKES

| | grams | **Weight:** | 35 | 50 |
|---|---|---|---|---|

40% bran* | with raisins*

| | | 40% bran* | with raisins* |
|---|---|---|---|
| % | **Water** | 3 | 7 |
| | **Calories** | 105 | 145 |
| grams | **Protein** | 4 | 4 |
| | **Fat** | 1 | 1 |
| | **Carbohydrate** | 28 | 40 |
| | **Calcium** | 19 | 28 |
| milligrams | **Phosphorus** | 125 | 146 |
| | **Iron** | 15.6 | 16.9 |
| | **Sodium** | **207 | **212 |
| | **Potassium** | 137 | 154 |
| | **Thiamin** | .41 | .58 |
| | **Riboflavin** | .49 | .71 |
| | **Niacin** | 4.1 | 5.8 |
| | **Ascorbic Acid** | 12 | 18 |
| unit | **Vitamin A** | 1,650 | 2,350 |

*Added sugar, salt, iron, vitamins.
**Based on revised value per 100g of product.

# CORN FLAKES
## PLAIN AND SUGARED

| grams | Weight: | 25 | 40 |
|---|---|---|---|

| | | plain*<br>added sugar | sugar<br>coated* |
|---|---|---|---|
| % | Water | 4 | 2 |
| | Calories | 95 | 155 |
| grams | Protein | 2 | 2 |
| | Fat | Trace | Trace |
| | Carbohydrate | 21 | 37 |
| | Calcium | ** | 1 |
| milligrams | Phosphorus | 9 | 10 |
| | Iron | .6 | 1.0 |
| | Sodium | 251 | 267 |
| | Potassium | 30 | 27 |
| | Thiamin | .29 | .46 |
| | Riboflavin | .35 | .56 |
| | Niacin | 2.9 | 4.6 |
| | Ascorbic Acid | 9 | 14 |
| unit | Vitamin A | 1,180 | 1,880 |

*Added salt, iron, vitamins.
**Value varies with the brand. Consult the label.

# PUFFED CORN*

| grams | **Weight:** | 20 |
|---|---|---|

*

| % | **Water** | 4 |
|---|---|---|
| | **Calories** | 80 |
| grams | **Protein** | 2 |
| | **Fat** | 1 |
| | **Carbohydrate** | 16 |
| | **Calcium** | 4 |
| milligrams | **Phosphorus** | 18 |
| | **Iron** | 2.3 |
| | **Sodium** | 233 |
| | **Potassium** | N.A. |
| | **Thiamin** | .23 |
| | **Riboflavin** | .28 |
| | **Niacin** | 2.3 |
| | **Ascorbic Acid** | 7 |
| unit | **Vitamin A** | 940 |

*Added sugar, salt, iron, vitamins.

# PUFFED OATS*

grams    **Weight:**       25

| | | |
|---|---|---|
| % | **Water** | 3 |
| | **Calories** | 100 |
| grams | **Protein** | 3 |
| | **Fat** | 1 |
| | **Carbohydrate** | 19 |
| | **Calcium** | 44 |
| milligrams | **Phosphorus** | 102 |
| | **Iron** | 2.9 |
| | **Sodium** | 317 |
| | **Potassium** | N.A. |
| | **Thiamin** | .29 |
| | **Riboflavin** | .35 |
| | **Niacin** | 2.9 |
| | **Ascorbic Acid** | 9 |
| unit | **Vitamin A** | 1,180 |

*Added sugar, salt, minerals, vitamins.

# PUFFED RICE
## PLAIN & PRE-SWEETENED

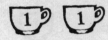

| | grams Weight: | 15 | 45 |
|---|---|---|---|
| | | plain* | pre-sweetened** |
| % | Water | 4 | 1.8 |
| | Calories | 60 | 175 |
| grams | Protein | 1 | 1.9 |
| | Fat | Trace | .3 |
| | Carbohydrate | 13 | 40.8 |
| | Calcium | 3 | 16 |
| | Phosphorus | 14 | 33 |
| | Iron | .3 | ***.9 |
| milligrams | Sodium | Trace | 148 |
| | Potassium | 15 | 33 |
| | Thiamin | .07 | .41 |
| | Riboflavin | .01 | .49 |
| | Niacin | .7 | 4.1 |
| | Ascorbic Acid | 0 | ****12 |
| unit | Vitamin A | 0 | 1,650 |

*Added iron, thiamin, niacin, **salt, iron and vitamins.
***Value varies with the brand. Consult the label.
****Applies to product with added ascorbic acid.

421

# PUFFED WHEAT
## PLAIN & PRE-SWEETENED

| | grams | Weight: | 15 | 35 |
|---|---|---|---|---|
| | | | plain* | pre-sweetened** |

| | | plain* | pre-sweetened** |
|---|---|---|---|
| % | Water | 3 | 2.8 |
| grams | Calories | 55 | 132 |
| | Protein | 2 | 2.1 |
| | Fat | Trace | .7 |
| | Carbohydrate | 12 | 30.9 |
| | Calcium | 4 | 7 |
| milligrams | Phosphorus | 48 | 53 |
| | Iron | .6 | ***1.2 |
| | Sodium | 1 | 56 |
| | Potassium | 51 | 61 |
| | Thiamin | .08 | .41 |
| | Riboflavin | .03 | .49 |
| | Niacin | 1.2 | 4.1 |
| | Ascorbic Acid | 0 | 12 |
| unit | Vitamin A | 0 | ****1,650 |

*Added iron, thiamin, niacin.   **Added salt, iron, vitamins.
***Value varies with brand. Consult label.
****Applies to product with added ascorbic acid.

# SHREDDED CORN*

grams  **Weight:**  25

| | |
|---|---|
| % | |
| **Water** | 3 |
| **Calories** | 95 |
| **Protein** | 2 |
| **Fat** | Trace |
| **Carbohydrate** | 22 |
| **Calcium** | 1 |
| **Phosphorus** | 10 |
| **Iron** | .6 |
| **Sodium** | **269 |
| **Potassium** | N.A. |
| **Thiamin** | .11 |
| **Riboflavin** | .05 |
| **Niacin** | .5 |
| **Ascorbic Acid** | 0 |
| **Vitamin A** | 0 |

grams {  ... milligrams {  ... unit

*Added sugar, salt, iron, thiamin, niacin.
**Based on cornbread made with white cornmeal.

423

# SHREDDED WHEAT

1 oblong
biscuit or ½
cup spoon-
size biscuit

| grams | **Weight:** | 25 |
|---|---|---|

| | | |
|---|---|---|
| % | **Water** | 7 |
| | **Calories** | 90 |
| grams | **Protein** | 2 |
| | **Fat** | 1 |
| | **Carbohydrate** | 20 |
| | **Calcium** | 11 |
| milligrams | **Phosphorus** | 97 |
| | **Iron** | .9 |
| | **Sodium** | 1 |
| | **Potassium** | .87 |
| | **Thiamin** | .06 |
| | **Riboflavin** | .03 |
| | **Niacin** | 1.1 |
| | **Ascorbic Acid** | 0 |
| unit | **Vitamin A** | 0 |

# WHEATGERM
## TOASTED*

grams **Weight:** 6

| | | |
|---|---|---|
| % | **Water** | 4 |
| | **Calories** | 25 |
| grams | **Protein** | 2 |
| | **Fat** | 1 |
| | **Carbohydrate** | 3 |
| | **Calcium** | 3 |
| | **Phosphorus** | 70 |
| | **Iron** | 5 |
| milligrams | **Sodium** | Trace |
| | **Potassium** | 57 |
| | **Thiamin** | .11 |
| | **Riboflavin** | .05 |
| | **Niacin** | .3 |
| | **Ascorbic Acid** | 1 |
| unit | **Vitamin A** | 10 |

*Without salt and sugar.

425

# WHEAT FLAKES*

grams    **Weight:**    30

| | |
|---|---|
| %   **Water** | 4 |
| **Calories** | 105 |
| **Protein** | 3 |
| **Fat** | Trace |
| **Carbohydrate** | 24 |
| **Calcium** | 12 |
| **Phosphorus** | 83 |
| **Iron** | ** |
| **Sodium** | 310 |
| **Potassium** | 81 |
| **Thiamin** | .35 |
| **Riboflavin** | .42 |
| **Niacin** | 3.5 |
| **Ascorbic Acid** | 11 |
| unit ▸   **Vitamin A** | 1,410 |

*Added sugar, salt, iron, vitamins.
**Value varies with the brand. Consult the label.

# CORN GRITS
## (HOMINY)*

| | Weight: | 245 | 245 |
|---|---|---|---|
| grams | | | |

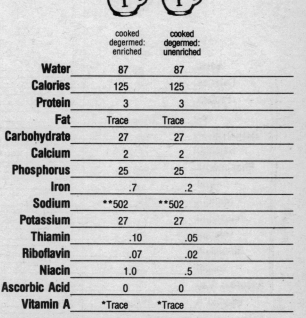

| | | cooked<br>degermed:<br>enriched | cooked<br>degermed:<br>unenriched |
|---|---|---|---|
| % | **Water** | 87 | 87 |
| | **Calories** | 125 | 125 |
| grams | **Protein** | 3 | 3 |
| | **Fat** | Trace | Trace |
| | **Carbohydrate** | 27 | 27 |
| | **Calcium** | 2 | 2 |
| milligrams | **Phosphorus** | 25 | 25 |
| | **Iron** | .7 | .2 |
| | **Sodium** | **502 | **502 |
| | **Potassium** | 27 | 27 |
| | **Thiamin** | .10 | .05 |
| | **Riboflavin** | .07 | .02 |
| | **Niacin** | 1.0 | .5 |
| | **Ascorbic Acid** | 0 | 0 |
| unit | **Vitamin A** | *Trace | *Trace |

*Applies to white varieties. For yellow varieties, value is 150 I.U.
**Based on value of 205 mg. per 100g for product cooked with salt added as specified by manufacturers. If cooked without added salt, value is negligible.

# FARINA
## (COOKED, HOT)

| | | |
|---|---|---|
| grams | **Weight:** | 245 |

quick
cooking,
enriched

| | | |
|---|---|---|
| % | **Water** | 89 |
| | **Calories** | 105 |
| grams | **Protein** | 3 |
| | **Fat** | Trace |
| | **Carbohydrate** | 22 |
| | **Calcium** | 147 |
| milligrams | **Phosphorus** | *113 |
| | **Iron** | ** |
| | **Sodium** | 353 |
| | **Potassium** | 25 |
| | **Thiamin** | .12 |
| | **Riboflavin** | .07 |
| | **Niacin** | 1.0 |
| | **Ascorbic Acid** | 0 |
| unit | **Vitamin A** | 0 |

*Applies to products that do not contain di-sodium phosphate. If di-sodium phosphate is an ingredient, value is 162 mg.
**Depends on brand. Could range from less than 1 mg-8 mg.

# OATMEAL OR ROLLED OATS

grams  **Weight:**  240

cooked, hot

| | | |
|---|---|---|
| % | **Water** | 87 |
| grams { | **Calories** | 130 |
| | **Protein** | 5 |
| | **Fat** | 2 |
| | **Carbohydrate** | 23 |
| | **Calcium** | 22 |
| milligrams { | **Phosphorus** | 137 |
| | **Iron** | 1.4 |
| | **Sodium** | *523 |
| | **Potassium** | 146 |
| | **Thiamin** | .19 |
| | **Riboflavin** | .05 |
| | **Niacin** | .2 |
| | **Ascorbic Acid** | 0 |
| unit | **Vitamin A** | 0 |

*Applies to product cooked with salt added as specified by manufacturers. If cooked without salt, value is negligible.

429

# ROLLED WHEAT

grams **Weight:** 240

cooked, hot

| | | |
|---|---|---|
| % | **Water** | 80 |
| grams { | **Calories** | 180 |
| | **Protein** | 5 |
| | **Fat** | 1 |
| | **Carbohydrate** | 41 |
| | **Calcium** | 19 |
| milligrams { | **Phosphorus** | 182 |
| | **Iron** | 1.7 |
| | **Sodium** | *708 |
| | **Potassium** | 202 |
| | **Thiamin** | .17 |
| | **Riboflavin** | .07 |
| | **Niacin** | 2.2 |
| | **Ascorbic Acid** | 0 |
| unit | **Vitamin A** | 0 |

*Based on revised value of 295 mg. per 100g for product cooked with salt added as specified by manufacturers.

430

# WHOLE WHEAT MEAL

grams **Weight:** 245

cooked, hot

| | % | |
|---|---|---|
| **Water** | 88 | |
| **Calories** | 110 | |

| grams { | | |
|---|---|---|
| **Protein** | 4 | |
| **Fat** | 1 | |
| **Carbohydrate** | 23 | |
| **Calcium** | 17 | |

| milligrams { | | |
|---|---|---|
| **Phosphorus** | 127 | |
| **Iron** | 1.2 | |
| **Sodium** | *519 | |
| **Potassium** | 118 | |
| **Thiamin** | .15 | |
| **Riboflavin** | .05 | |
| **Niacin** | 1.5 | |
| **Ascorbic Acid** | 0 | |
| **Vitamin A** | 0 | |

unit

*Applies to product cooked with salt added as specified by manufacturers. If cooked wtihout added salt, value is negligible.

# PANCAKES*

| grams | **Weight:** | 27 | 27 | 27 |
|---|---|---|---|---|
| | | 1 | 1 | 1 |
| | | buckwheat, made from mix** | plain: from home recipe using enriched flour | plain: from mix*** |
| % | **Water** | 58 | 50 | 51 |
| | **Calories** | 55 | 60 | 60 |
| grams | **Protein** | 2 | 2 | 2 |
| | **Fat** | 2 | 2 | 2 |
| | **Carbohydrate** | 6 | 9 | 9 |
| | **Calcium** | 59 | 27 | 58 |
| milligrams | **Phosphorus** | 91 | 38 | 70 |
| | **Iron** | .4 | .4 | .3 |
| | **Sodium** | 125 | 152 | 152 |
| | **Potassium** | 66 | 33 | 42 |
| | **Thiamin** | .04 | .06 | .04 |
| | **Riboflavin** | .05 | .07 | .06 |
| | **Niacin** | .2 | .5 | .2 |
| | **Ascorbic Acid** | Trace | Trace | Trace |
| unit | **Vitamin A** | 60 | 30 | 70 |

*Made with vegetable shortening.
**Made with buckwheat and enriched flours, egg and milk added.
***Made with enriched flour, egg and milk added.

# WAFFLES*

| | from home recipe | from mix, egg & milk added |
|---|---|---|
| **Weight:** grams | 75 | 75 |
| **Water** % | 41 | 42 |
| **Calories** | 210 | 205 |
| **Protein** grams | 7 | 7 |
| **Fat** | 7 | 8 |
| **Carbohydrate** | 28 | 27 |
| **Calcium** | 85 | 179 |
| **Phosphorus** | 130 | 257 |
| **Iron** | 1.3 | 1.0 |
| **Sodium** milligrams | 356 | 515 |
| **Potassium** | 109 | 146 |
| **Thiamin** | .17 | .14 |
| **Riboflavin** | .23 | .22 |
| **Niacin** | 1.4 | .9 |
| **Ascorbic Acid** | Trace | Trace |
| **Vitamin A** unit | 250 | 170 |

*Made with enriched flour and vegetable shortening.

# LEGUMES

BLACKEYE PEAS
CHICKPEAS OR GARBONZOS
GREAT NORTHERN BEANS
LENTILS
LIMA BEANS
NAVY OR PEA BEANS
RED OR KIDNEY BEANS
SPLIT PEAS
WHITE BEANS

# BLACKEYE PEAS

grams **Weight:** 250

cooked (with
residual
cooking
liquid)

| | | |
|---|---|---|
| % | **Water** | 80 |
| | **Calories** | 190 |
| grams { | **Protein** | 13 |
| | **Fat** | 1 |
| | **Carbohydrate** | 35 |
| | **Calcium** | 43 |
| milligrams { | **Phosphorus** | 238 |
| | **Iron** | 33. |
| | **Sodium** | *2 |
| | **Potassium** | 573 |
| | **Thiamin** | .40 |
| | **Riboflavin** | .10 |
| | **Niacin** | 1.0 |
| | **Ascorbic Acid** | N.A. |
| unit | **Vitamin A** | 30 |

*Value for product without added salt.

# CHICKPEAS
## OR GARBONZOS

grams **Weight:** 200

dry, raw

| | |
|---|---|
| % | Water | 10.7 |

| | | |
|---|---|---|
| % | Water | 10.7 |
| grams | Calories | 720 |
| | Protein | 41 |
| | Fat | 9.6 |
| | Carbohydrate | 122 |
| | Calcium | 300 |
| milligrams | Phosphorus | 662 |
| | Iron | 13.8 |
| | Sodium | 52 |
| | Potassium | 1,594 |
| | Thiamin | .62 |
| | Riboflavin | .30 |
| | Niacin | 4 |
| | Ascorbic Acid | N.A. |
| unit | Vitamin A | 100 |

# GREAT NORTHERN BEANS

| grams | Weight: | 180 |
| --- | --- | --- |

cooked,
drained

| | | |
| --- | --- | --- |
| % | **Water** | 69 |
| | **Calories** | 210 |
| grams { | **Protein** | 14 |
| | **Fat** | 1 |
| | **Carbohydrate** | 38 |
| | **Calcium** | 90 |
| | **Phosphorus** | 266 |
| | **Iron** | 4.8 |
| | **Sodium** | 34 |
| milligrams { | **Potassium** | 749 |
| | **Thiamin** | .25 |
| | **Riboflavin** | .13 |
| | **Niacin** | 1.3 |
| | **Ascorbic Acid** | 0 |
| unit | **Vitamin A** | 0 |

# LENTILS

grams  **Weight:**  200

cooked

| | | |
|---|---|---|
| % | **Water** | 72 |
| | **Calories** | 210 |
| | **Protein** | 16 |
| grams | **Fat** | Trace |
| | **Carbohydrate** | 39 |
| | **Calcium** | 50 |
| | **Phosphorus** | 238 |
| | **Iron** | 4.2 |
| | **Sodium** | *N.A. |
| milligrams | **Potassium** | 498 |
| | **Thiamin** | .14 |
| | **Riboflavin** | .12 |
| | **Niacin** | 1.2 |
| | **Ascorbic Acid** | 0 |
| unit | **Vitamin A** | 40 |

*Value for product without added salt.

438

# LIMA BEANS

grams **Weight:** 190

cooked,
drained

| | | |
|---|---|---|
| % | Water | 64 |
| | Calories | 260 |
| grams | Protein | 16 |
| | Fat | 1 |
| | Carbohydrate | 49 |
| | Calcium | 55 |
| | Phosphorus | 293 |
| | Iron | 5.9 |
| milligrams | Sodium | *2 |
| | Potassium | 1,163 |
| | Thiamin | .25 |
| | Riboflavin | .11 |
| | Niacin | 1.3 |
| | Ascorbic Acid | N.A. |
| unit | Vitamin A | N.A. |

*Value is for unsalted product. If salt is used, increase value by 236 per 100g of vegetable—estimated typical amount of salt (0.6%) in canned vegetables.

# NAVY OR PEA BEANS

| | grams | **Weight:** | 190 |
|---|---|---|---|

cooked,
drained

| % | **Water** | 69 |
|---|---|---|
| | **Calories** | 225 |
| grams { | **Protein** | 15 |
| | **Fat** | 1 |
| | **Carbohydrate** | 40 |
| | **Calcium** | 95 |
| | **Phosphorus** | 281 |
| | **Iron** | 5.1 |
| milligrams { | **Sodium** | 39 |
| | **Potassium** | 790 |
| | **Thiamin** | .27 |
| | **Riboflavin** | .13 |
| | **Niacin** | 1.3 |
| | **Ascorbic Acid** | 0 |
| unit | **Vitamin A** | 0 |

# RED OR KIDNEY BEANS

grams **Weight:** 255

canned solids
& liquid

| | | |
|---|---|---|
| % | **Water** | 76 |
| | **Calories** | 230 |
| grams | **Protein** | 15 |
| | **Fat** | 1 |
| | **Carbohydrate** | 42 |
| | **Calcium** | 74 |
| | **Phosphorus** | 278 |
| | **Iron** | 4.6 |
| milligrams | **Sodium** | *8 |
| | **Potassium** | 673 |
| | **Thiamin** | .13 |
| | **Riboflavin** | .10 |
| | **Niacin** | 1.5 |
| | **Ascorbic Acid** | N.A. |
| unit | **Vitamin A** | 10 |

*Value for product without added salt.

# SPLIT PEAS

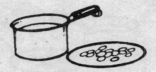

grams | **Weight:** | 200

dry, cooked

| % | Water | 70 |
|---|---|---|
| | **Calories** | 230 |
| | **Protein** | 16 |
| grams | **Fat** | 1 |
| | **Carbohydrate** | 42 |
| | **Calcium** | 22 |
| | **Phosphorus** | 178 |
| | **Iron** | 3.4 |
| | **Sodium** | *26 |
| | **Potassium** | 592 |
| milligrams | **Thiamin** | .30 |
| | **Riboflavin** | .18 |
| | **Niacin** | 1.8 |
| | **Ascorbic Acid** | N.A. |
| unit | **Vitamin A** | 80 |

*Value for product without added salt.

# WHITE BEANS

| | grams | Weight: | 255 | 255 | 255 |
|---|---|---|---|---|---|

| | | frankfurters | with pork & tomato sauce | pork & sweet sauce |
|---|---|---|---|---|
| % | **Water** | 71 | 71 | 66 |
| | **Calories** | 365 | 310 | 385 |
| grams | **Protein** | 19 | 16 | 16 |
| | **Fat** | 18 | 7 | 12 |
| | **Carbohydrate** | 32 | 48 | 54 |
| | **Calcium** | 94 | 138 | 161 |
| milligrams | **Phosphorus** | 303 | 235 | 291 |
| | **Iron** | 4.8 | 4.6 | 5.9 |
| | **Sodium** | N.A. | 1,181 | 969 |
| | **Potassium** | 668 | 536 | N.A. |
| | **Thiamin** | .18 | .20 | .15 |
| | **Riboflavin** | .15 | .08 | .10 |
| | **Niacin** | 3.3 | 1.5 | 1.3 |
| | **Ascorbic Acid** | Trace | 5 | N.A. |
| unit | **Vitamin A** | 330 | 330 | N.A. |

Canned solids and liquid.

# SEEDS AND NUTS

ALMONDS
BEECHNUTS
BRAZIL NUTS
BUTTERNUTS
CASHEWS
CHESTNUTS
COCONUT MEAT
HAZELNUTS (FILBERTS)
PEANUTS
PEANUT BUTTER
PECANS
PUMPKIN OR SQUASH KERNELS
SUNFLOWER SEEDS
WALNUTS

# ALMONDS*

| | | |
|---|---|---|
| grams | **Weight:** | 130 | 115 |

| | 1 | 1 |
|---|---|---|
| | chopped, about 130 almonds | slivered, not pressed down, about 115 almonds |

| | | | |
|---|---|---:|---:|
| % | **Water** | 5 | 5 |
| | **Calories** | 775 | 690 |
| | **Protein** | 24 | 21 |
| grams | **Fat** | 70 | 62 |
| | **Carbohydrate** | 25 | 22 |
| | **Calcium** | 304 | 269 |
| | **Phosphorus** | 655 | 580 |
| | **Iron** | 6.1 | 5.4 |
| | **Sodium** | 5 | 5 |
| milligrams | **Potassium** | 1,005 | 889 |
| | **Thiamin** | .31 | .28 |
| | **Riboflavin** | 1.20 | 1.06 |
| | **Niacin** | 4.6 | 4.0 |
| | **Ascorbic Acid** | Trace | Trace |
| unit | **Vitamin A** | 0 | 0 |

*Shelled.

445

# BEECHNUTS

grams    **Weight:**      454

shelled

| | | |
|---|---|---|
| % | **Water** | 6.6 |
| | **Calories** | 2,576 |
| grams | **Protein** | 88 |
| | **Fat** | 226.8 |
| | **Carbohydrate** | 92.1 |
| | **Calcium** | N.A. |
| | **Phosphorus** | N.A. |
| | **Iron** | N.A. |
| | **Sodium** | N.A. |
| milligrams | **Potassium** | N.A. |
| | **Thiamin** | N.A. |
| | **Riboflavin** | N.A. |
| | **Niacin** | N.A. |
| | **Ascorbic Acid** | N.A. |
| unit | **Vitamin A** | N.A. |

# BRAZIL NUTS

grams    **Weight:**      28

shelled (6-8
large kernels)

| | | |
|---|---|---|
| % | **Water** | 5 |
| | **Calories** | 185 |
| grams | **Protein** | 4 |
| | **Fat** | 19 |
| | **Carbohydrate** | 3 |
| | **Calcium** | 53 |
| | **Phosphorus** | 196 |
| | **Iron** | 1.0 |
| | **Sodium** | Trace |
| milligrams | **Potassium** | 203 |
| | **Thiamin** | .27 |
| | **Riboflavin** | .03 |
| | **Niacin** | .5 |
| | **Ascorbic Acid** | N.A. |
| unit | **Vitamin A** | Trace |

447

# BUTTERNUTS

grams  **Weight:**      454

shelled

| | | |
|---|---|---|
| % | **Water** | 3.8 |
| | **Calories** | 2,853 |
| grams { | **Protein** | 107.5 |
| | **Fat** | 277.6 |
| | **Carbohydrate** | 38.1 |
| | **Calcium** | N.A. |
| | **Phosphorus** | N.A. |
| | **Iron** | N.A. |
| | **Sodium** | N.A. |
| milligrams { | **Potassium** | N.A. |
| | **Thiamin** | N.A. |
| | **Riboflavin** | N.A. |
| | **Niacin** | N.A. |
| | **Ascorbic Acid** | N.A. |
| unit | **Vitamin A** | N.A. |

# CASHEWS

Weight: 140 grams

roasted in oil

| | | |
|---|---|---|
| % | Water | 5 |
| | Calories | 785 |
| grams | Protein | 24 |
| | Fat | 64 |
| | Carbohydrate | 41 |
| | Calcium | 53 |
| | Phosphorus | 522 |
| | Iron | 5.3 |
| milligrams | Sodium | *21 |
| | Potassium | 650 |
| | Thiamin | .60 |
| | Riboflavin | .35 |
| | Niacin | 2.5 |
| | Ascorbic Acid | N.A. |
| unit | Vitamin A | 140 |

*Applies to unsalted nuts. For salted nuts, value for 1 cup of kernels is approximately 280 mg.

# CHESTNUTS

grams **Weight:** 160

shelled

| | |
|---|---|
| % Water | 52.5 |
| Calories | 310 |
| Protein | 4.6 |
| Fat | 2.4 |
| Carbohydrate | 67.4 |
| Calcium | 43 |
| Phosphorus | 141 |
| Iron | 2.7 |
| Sodium | 10 |
| Potassium | 726 |
| Thiamin | .35 |
| Riboflavin | .35 |
| Niacin | 1 |
| Ascorbic Acid | N.A. |
| Vitamin A | N.A. |

grams { Protein, Fat, Carbohydrate

milligrams { Phosphorus, Iron, Sodium, Potassium, Thiamin, Riboflavin, Niacin

unit

# COCONUT MEAT
## FRESH

| | grams | Weight: | 45 | 80 | |
|---|---|---|---|---|---|

piece about 2 x 2 x ½"    shredded or grated, not pressed down

| | | 45 | 80 |
|---|---|---|---|
| % | Water | 51 | 51 |
| | Calories | 155 | 275 |
| grams | Protein | 2 | 3 |
| | Fat | 16 | 28 |
| | Carbohydrate | 4 | 8 |
| | Calcium | 6 | 10 |
| milligrams | Phosphorus | 43 | 76 |
| | Iron | .8 | 1.4 |
| | Sodium | 10 | 18 |
| | Potassium | 115 | 205 |
| | Thiamin | .02 | .04 |
| | Riboflavin | .01 | .02 |
| | Niacin | .2 | .4 |
| | Ascorbic Acid | 1 | 2 |
| unit | Vitamin A | 0 | 0 |

# HAZELNUTS
## (FILBERTS)

grams **Weight:** 115

80 kernels,
chopped

| | | |
|---|---|---|
| % | **Water** | 6 |
| | **Calories** | 730 |
| grams { | **Protein** | 14 |
| | **Fat** | 72 |
| | **Carbohydrate** | 19 |
| | **Calcium** | 240 |
| | **Phosphorus** | 388 |
| | **Iron** | 3.9 |
| | **Sodium** | 2 |
| milligrams { | **Potassium** | 810 |
| | **Thiamin** | .53 |
| | **Riboflavin** | N.A. |
| | **Niacin** | 1.0 |
| | **Ascorbic Acid** | Trace |
| unit | **Vitamin A** | N.A. |

# PEANUTS*

| | | |
|---|---|---|
| grams | **Weight:** | 144 |

roasted in oil,
salted
(whole, halves,
chopped)

| | | |
|---|---|---|
| % | **Water** | 2 |
| | **Calories** | 840 |
| grams | **Protein** | 37 |
| | **Fat** | 72 |
| | **Carbohydrate** | 27 |
| | **Calcium** | 107 |
| milligrams | **Phosphorus** | 577 |
| | **Iron** | 3.0 |
| | **Sodium** | 602 |
| | **Potassium** | 971 |
| | **Thiamin** | .46 |
| | **Riboflavin** | .19 |
| | **Niacin** | 24.8 |
| | **Ascorbic Acid** | 0 |
| unit | **Vitamin A** | N.A. |

*Shelled, coated, salted.

# PEANUT BUTTER

grams    **Weight:**    16

*

| | | |
|---|---|---|
| % | **Water** | 2 |
| | **Calories** | 95 |
| grams | **Protein** | 4 |
| | **Fat** | 8 |
| | **Carbohydrate** | 3 |
| | **Calcium** | 9 |
| milligrams | **Phosphorus** | 61 |
| | **Iron** | .3 |
| | **Sodium** | 97 |
| | **Potassium** | 100 |
| | **Thiamin** | .02 |
| | **Riboflavin** | .02 |
| | **Niacin** | 2.4 |
| | **Ascorbic Acid** | 0 |
| unit | **Vitamin A** | N.A. |

*Made with moderate amounts of added fat, nutritive sweetener, salt.

# PECANS

grams **Weight:** 118

shelled,
chopped,
about 120
large kernels

| | | |
|---|---|---|
| % | **Water** | 3 |
| | **Calories** | 810 |
| grams | **Protein** | 11 |
| | **Fat** | 84 |
| | **Carbohydrate** | 17 |
| | **Calcium** | 86 |
| | **Phosphorus** | 341 |
| | **Iron** | 2.8 |
| | **Sodium** | Trace |
| milligrams | **Potassium** | 712 |
| | **Thiamin** | 1.01 |
| | **Riboflavin** | .15 |
| | **Niacin** | 1.1 |
| | **Ascorbic Acid** | 2 |
| unit | **Vitamin A** | 150 |

# PUMPKIN OR SQUASH
## KERNELS

grams **Weight:** 140

dry, hulled

| | | |
|---|---|---|
| % **Water** | 4 | |
| **Calories** | 775 | |
| grams **Protein** | 41 | |
| **Fat** | 65 | |
| **Carbohydrate** | 21 | |
| **Calcium** | 71 | |
| **Phosphorus** | 1,602 | |
| **Iron** | 15.7 | |
| **Sodium** | N.A. | |
| milligrams **Potassium** | 1,386 | |
| **Thiamin** | .34 | |
| **Riboflavin** | .27 | |
| **Niacin** | 3.4 | |
| **Ascorbic Acid** | N.A. | |
| unit **Vitamin A** | 100 | |

# SUNFLOWER SEEDS

grams **Weight:** 145

dry, hulled

| | | |
|---|---|---|
| % | **Water** | 5 |
| | **Calories** | 810 |
| grams | **Protein** | 35 |
| | **Fat** | 69 |
| | **Carbohydrate** | 29 |
| | **Calcium** | 174 |
| milligrams | **Phosphorus** | 1,214 |
| | **Iron** | 10.3 |
| | **Sodium** | 44 |
| | **Potassium** | 1,334 |
| | **Thiamin** | 2.84 |
| | **Riboflavin** | .33 |
| | **Niacin** | 7.8 |
| | **Ascorbic Acid** | N.A. |
| unit | **Vitamin A** | 70 |

# WALNUTS
## BLACK AND PERSIAN

| grams | **Weight:** | 125 | 80 | 120 |
|---|---|---|---|---|
| | |  | |  |
| | | black & shelled, chopped or broken kernels | black & shelled, ground (finely) | Persian or English, shelled, chopped (about 60 halves) |
| % | **Water** | 3 | 3 | 4 |
| grams | **Calories** | 785 | 500 | 780 |
| | **Protein** | 26 | 16 | 18 |
| | **Fat** | 74 | 47 | 77 |
| | **Carbohydrate** | 19 | 12 | 19 |
| milligrams | **Calcium** | Trace | Trace | 119 |
| | **Phosphorus** | 713 | 456 | 456 |
| | **Iron** | 7.5 | 4.8 | 3.7 |
| | **Sodium** | 4 | 2 | 2 |
| | **Potassium** | 515 | 368 | 540 |
| | **Thiamin** | .28 | .18 | .40 |
| | **Riboflavin** | .14 | .09 | .16 |
| | **Niacin** | .9 | .6 | 1.1 |
| | **Ascorbic Acid** | N.A. | N.A. | 2 |
| unit | **Vitamin A** | 380 | 240 | 40 |

# CANDIES

BUTTERSCOTCH
CARAMELS
CHEWING GUM, CANDY COATED
CHOCOLATE, BITTER OR BAKING
CHOCOLATE-COVERED COCONUT BAR
CHOCOLATE-COVERED PEANUTS
CHOCOLATE-FLAVORED ROLL
CHOCOLATE FUDGE
CHOCOLATE, MILK
CHOCOLATE, SEMI-SWEET
FONDANT (UNCOATED MINTS, CANDY CORN)
GUM DROPS
HARD CANDY
JELLY BEANS
MARSHMALLOWS

# BUTTERSCOTCH

**Weight:** 28
grams

| | | |
|---|---|---|
| % | **Water** | 1.5 |
| | **Calories** | 113 |
| grams { | **Protein** | Trace |
| | **Fat** | 1 |
| | **Carbohydrate** | 26.9 |
| | **Calcium** | 5 |
| milligrams { | **Phosphorus** | 2 |
| | **Iron** | .4 |
| | **Sodium** | 19 |
| | **Potassium** | 1 |
| | **Thiamin** | 0 |
| | **Riboflavin** | Trace |
| | **Niacin** | Trace |
| | **Ascorbic Acid** | 0 |
| unit | **Vitamin A** | 40 |

# CARAMELS
## PLAIN OR CHOCOLATE

| | grams | Weight: | 28 | 28 |
|---|---|---|---|---|

| | | plain or chocolate | plain or chocolate with nuts |
|---|---|---|---|
| % | **Water** | 8 | 7.1 |
| | **Calories** | 115 | 121 |
| grams | **Protein** | 2 | 1.3 |
| | **Fat** | 3 | 4.6 |
| | **Carbohydrate** | 22 | 20 |
| | **Calcium** | 42 | 40 |
| milligrams | **Phosphorus** | 35 | 39 |
| | **Iron** | .4 | .4 |
| | **Sodium** | 64 | 58 |
| | **Potassium** | 54 | 66 |
| | **Thiamin** | .01 | .03 |
| | **Riboflavin** | .05 | .05 |
| | **Niacin** | .1 | .1 |
| | **Ascorbic Acid** | Trace | Trace |
| unit | **Vitamin A** | Trace | 10 |

# CHEWING GUM
## CANDY COATED PIECES

| grams | **Weight:** | 1.7 |
|---|---|---|

1

3¼" x ½"
x ¼"

| | | |
|---|---|---|
| % | **Water** | 3.5 |
| | **Calories** | 5 |
| grams | **Protein** | N.A. |
| | **Fat** | N.A. |
| | **Carbohydrate** | 1.6 |
| | **Calcium** | N.A. |
| | **Phosphorus** | N.A. |
| | **Iron** | N.A. |
| | **Sodium** | N.A. |
| milligrams | **Potassium** | N.A. |
| | **Thiamin** | 0 |
| | **Riboflavin** | 0 |
| | **Niacin** | 0 |
| | **Ascorbic Acid** | 0 |
| unit | **Vitamin A** | N.A. |

# CHOCOLATE
## BITTER OR BAKING

grams    **Weight:**      28

| | | |
|---|---|---|
| % | **Water** | 1.8 |
| | **Calories** | 1.35 |
| grams | **Protein** | 2.2 |
| | **Fat** | 11.3 |
| | **Carbohydrate** | 13.3 |
| | **Calcium** | 16 |
| | **Phosphorus** | 81 |
| | **Iron** | 1.4 |
| | **Sodium** | 1 |
| milligrams | **Potassium** | 174 |
| | **Thiamin** | .01 |
| | **Riboflavin** | .05 |
| | **Niacin** | .3 |
| | **Ascorbic Acid** | 0 |
| unit | **Vitamin A** | 10 |

# CHOCOLATE-COVERED
# COCONUT BAR

grams **Weight:** 28

| | | |
|---|---|---|
| % | **Water** | 6.6 |
| | **Calories** | 124 |
| grams | **Protein** | .8 |
| | **Fat** | 5 |
| | **Carbohydrate** | 20.4 |
| | **Calcium** | 14 |
| milligrams | **Phosphorus** | 22 |
| | **Iron** | .3 |
| | **Sodium** | 56 |
| | **Potassium** | 47 |
| | **Thiamin** | .01 |
| | **Riboflavin** | .02 |
| | **Niacin** | .1 |
| | **Ascorbic Acid** | 0 |
| unit | **Vitamin A** | 0 |

# CHOCOLATE-COVERED
# PEANUTS

grams **Weight:** 28

| % | Water | 1 | |
|---|---|---|---|
| | **Calories** | 160 | |
| grams | **Protein** | 5 | |
| | **Fat** | 12 | |
| | **Carbohydrate** | 11 | |
| | **Calcium** | 33 | |
| | **Phosphorus** | 84 | |
| | **Iron** | .4 | |
| milligrams | **Sodium** | 17 | |
| | **Potassium** | 143 | |
| | **Thiamin** | .10 | |
| | **Riboflavin** | .05 | |
| | **Niacin** | 2.1 | |
| | **Ascorbic Acid** | Trace | |
| unit | **Vitamin A** | Trace | |

# CHOCOLATE-FLAVORED
## ROLL

grams    **Weight:**              5

| | | |
|---|---|---|
| % | **Water** | 5.6 |
| | **Calories** | 20 |
| | **Protein** | .1 |
| | **Fat** | .4 |
| | **Carbohydrate** | 4.1 |
| | **Calcium** | 3 |
| | **Phosphorus** | 6 |
| | **Iron** | .1 |
| | **Sodium** | 10 |
| | **Potassium** | 6 |
| | **Thiamin** | Trace |
| | **Riboflavin** | Trace |
| | **Niacin** | Trace |
| | **Ascorbic Acid** | Trace |
| unit | **Vitamin A** | Trace |

(grams bracket: Calories, Protein, Fat, Carbohydrate, Calcium)
(milligrams bracket: Phosphorus, Iron, Sodium, Potassium, Thiamin, Riboflavin, Niacin, Ascorbic Acid)

# CHOCOLATE FUDGE

grams    **Weight:**     28

plain

| | | |
|---|---|---|
| % | **Water** | 8 |
| | **Calories** | 115 |
| grams | **Protein** | 1 |
| | **Fat** | 3 |
| | **Carbohydrate** | 21 |
| | **Calcium** | 22 |
| | **Phosphorus** | 24 |
| | **Iron** | .3 |
| milligrams | **Sodium** | 54 |
| | **Potassium** | 42 |
| | **Thiamin** | .01 |
| | **Riboflavin** | .03 |
| | **Niacin** | .1 |
| | **Ascorbic Acid** | Trace |
| unit | **Vitamin A** | Trace |

# CHOCOLATE MILK

grams  **Weight:**  28

plain

| % | Water | 1 | |
|---|---|---|---|
| | **Calories** | 145 | |
| grams | **Protein** | 2 | |
| | **Fat** | 9 | |
| | **Carbohydrate** | 16 | |
| | **Calcium** | 65 | |
| milligrams | **Phosphorus** | 65 | |
| | **Iron** | .3 | |
| | **Sodium** | 27 | |
| | **Potassium** | 109 | |
| | **Thiamin** | .02 | |
| | **Riboflavin** | .10 | |
| | **Niacin** | .1 | |
| | **Ascorbic Acid** | Trace | |
| unit | **Vitamin A** | 80 | |

# CHOCOLATE
## SEMI-SWEET

| grams | **Weight:** | 170 |
| --- | --- | --- |

pieces
60 per oz.*

| % | **Water** | 1 |
| --- | --- | --- |
| | **Calories** | 860 |
| grams { | **Protein** | 7 |
| | **Fat** | 61 |
| | **Carbohydrate** | 97 |
| | **Calcium** | 51 |
| milligrams { | **Phosphorus** | 255 |
| | **Iron** | 4.4 |
| | **Sodium** | 3 |
| | **Potassium** | 553 |
| | **Thiamin** | .02 |
| | **Riboflavin** | .14 |
| | **Niacin** | .9 |
| | **Ascorbic Acid** | 0 |
| unit | **Vitamin A** | 30 |

*or, 6 oz. pkg. approx.

469

# UNCOATED
# FONDANT
## (MINTS, CANDY CORN, ETC.)

grams    **Weight:**          28

| | | |
|---|---|---|
| % | **Water** | 8 |
| | **Calories** | 105 |
| grams | **Protein** | Trace |
| | **Fat** | 1 |
| | **Carbohydrate** | 25 |
| | **Calcium** | 4 |
| milligrams | **Phosphorus** | 2 |
| | **Iron** | .3 |
| | **Sodium** | 60 |
| | **Potassium** | 1 |
| | **Thiamin** | Trace |
| | **Riboflavin** | Trace |
| | **Niacin** | Trace |
| | **Ascorbic Acid** | 0 |
| unit | **Vitamin A** | 0 |

# GUM DROPS

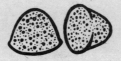

grams    **Weight:**    28

| | | |
|---|---|---|
| % | **Water** | 12 |
| | **Calories** | 100 |
| grams { | **Protein** | Trace |
| | **Fat** | Trace |
| | **Carbohydrate** | 25 |
| | **Calcium** | 2 |
| milligrams { | **Phosphorus** | Trace |
| | **Iron** | .1 |
| | **Sodium** | 10 |
| | **Potassium** | 1 |
| | **Thiamin** | 0 |
| | **Riboflavin** | Trace |
| | **Niacin** | Trace |
| | **Ascorbic Acid** | 0 |
| unit | **Vitamin A** | 0 |

# HARD CANDY

grams   **Weight:**          28

| | | |
|---|---|---|
| % | **Water** | 1 |
| | **Calories** | 110 |
| grams | **Protein** | 0 |
| | **Fat** | Trace |
| | **Carbohydrate** | 28 |
| | **Calcium** | 6 |
| milligrams | **Phosphorus** | 2 |
| | **Iron** | .5 |
| | **Sodium** | 9 |
| | **Potassium** | 1 |
| | **Thiamin** | 0 |
| | **Riboflavin** | 0 |
| | **Niacin** | 0 |
| | **Ascorbic Acid** | 0 |
| unit | **Vitamin A** | 0 |

# JELLYBEANS

grams **Weight:** 28

| | | |
|---|---|---|
| % | **Water** | 6.3 |
| | **Calories** | 104 |
| grams { | **Protein** | Trace |
| | **Fat** | .1 |
| | **Carbohydrate** | 26.4 |
| | **Calcium** | 3 |
| milligrams { | **Phosphorus** | 1 |
| | **Iron** | .3 |
| | **Sodium** | 3 |
| | **Potassium** | Trace |
| | **Thiamin** | 0 |
| | **Riboflavin** | Trace |
| | **Niacin** | Trace |
| | **Ascorbic Acid** | 0 |
| unit | **Vitamin A** | 0 |

*Approximately 10 jelly beans per one oz.

# MARSHMALLOWS

**Weight:** grams 28

| | | |
|---|---|---|
| % | **Water** | 17 |
| | **Calories** | 90 |
| grams | **Protein** | 1 |
| | **Fat** | Trace |
| | **Carbohydrate** | 23 |
| | **Calcium** | 5 |
| | **Phosphorus** | 2 |
| | **Iron** | .5 |
| milligrams | **Sodium** | 11 |
| | **Potassium** | 2 |
| | **Thiamin** | 0 |
| | **Riboflavin** | Trace |
| | **Niacin** | Trace |
| | **Ascorbic Acid** | 0 |
| unit | **Vitamin A** | 0 |

# SWEETENERS

BROWN SUGAR
CANE SYRUP OR MOLASSES
CAROB
CHOCOLATE-FLAVORED SYRUP
HONEY
SORGHUM
TABLE SYRUP BLENDS (CHIEFLY CORN)
WHITE SUGAR, GRANULATED
WHITE SUGAR, POWDERED

# BROWN SUGAR

grams | **Weight:** | 220

pressed
down

| | | |
|---|---|---|
| % | **Water** | 2 |
| | **Calories** | 820 |
| grams | **Protein** | 0 |
| | **Fat** | 0 |
| | **Carbohydrate** | 212 |
| | **Calcium** | 187 |
| milligrams | **Phosphorus** | 42 |
| | **Iron** | 7.5 |
| | **Sodium** | 66 |
| | **Potassium** | 757 |
| | **Thiamin** | .02 |
| | **Riboflavin** | .07 |
| | **Niacin** | .4 |
| | **Ascorbic Acid** | 0 |
| unit | **Vitamin A** | 0 |

# CANE SYRUP OR
# MOLASSES

| | grams | | |
|---|---|---|---|
| **Weight:** | | 20 | 20 |
| | | Light (first extraction) | Blackstrap (third extraction) |
| % | **Water** | 24 | 24 |
| | **Calories** | 50 | 45 |
| grams { | **Protein** | — | — |
| | **Fat** | — | — |
| | **Carbohydrate** | 13 | 11 |
| | **Calcium** | 33 | 137 |
| milligrams { | **Phosphorus** | 9 | 17 |
| | **Iron** | .9 | 3.2 |
| | **Sodium** | 3 | 19 |
| | **Potassium** | 183 | 585 |
| | **Thiamin** | .01 | .02 |
| | **Riboflavin** | .01 | .04 |
| | **Niacin** | Trace | .4 |
| | **Ascorbic Acid** | N.A. | N.A. |
| unit | **Vitamin A** | N.A. | N.A. |

# CAROB

| | grams | Weight: | 140 | 8 |
|---|---|---|---|---|

| % | | | | |
|---|---|---|---|---|
| | **Water** | 11.2 | 11.2 | |
| | **Calories** | 252 | 14 | |
| | **Protein** | 6.3 | .4 | |
| grams | **Fat** | 2 | .1 | |
| | **Carbohydrate** | 113 | 6.5 | |
| | **Calcium** | 493 | 28 | |
| | **Phosphorus** | 113 | 6 | |
| | **Iron** | N.A. | N.A. | |
| | **Sodium** | N.A. | N.A. | |
| milligrams | **Potassium** | N.A. | N.A. | |
| | **Thiamin** | N.A. | N.A. | |
| | **Riboflavin** | N.A. | N.A. | |
| | **Niacin** | N.A. | N.A. | |
| | **Ascorbic Acid** | N.A. | N.A. | |
| unit | **Vitamin A** | N.A. | N.A. | |

# CHOCOLATE-FLAVORED
# SYRUP
## THIN AND FUDGE TYPE

| grams | **Weight:** | 38 | 38 |
|---|---|---|---|

|  | thin type | fudge type |
|---|---|---|

| | | thin type | fudge type |
|---|---|---|---|
| % | **Water** | 32 | 25 |
| | **Calories** | 90 | 125 |
| grams { | **Protein** | 1 | 2 |
| | **Fat** | 1 | 5 |
| | **Carbohydrate** | 24 | 20 |
| | **Calcium** | 6 | 48 |
| milligrams { | **Phosphorus** | 35 | 60 |
| | **Iron** | .6 | .5 |
| | **Sodium** | 20 | 33 |
| | **Potassium** | 106 | 107 |
| | **Thiamin** | .01 | .02 |
| | **Riboflavin** | .03 | .08 |
| | **Niacin** | .2 | .2 |
| | **Ascorbic Acid** | 0 | Trace |
| unit | **Vitamin A** | Trace | 60 |

One fluid ounce.

479

# HONEY

grams | **Weight:** | 21

strained or
extracted

| % | Water | 17 |
|---|---|---|
| grams { | Calories | 65 |
| | Protein | Trace |
| | Fat | 0 |
| | Carbohydrate | 17 |
| | Calcium | 1 |
| milligrams { | Phosphorus | 1 |
| | Iron | .1 |
| | Sodium | 1 |
| | Potassium | 11 |
| | Thiamin | Trace |
| | Riboflavin | .01 |
| | Niacin | .1 |
| | Ascorbic Acid | Trace |
| unit | Vitamin A | 0 |

# SORGHUM

grams **Weight:** 21

| | | |
|---|---|---|
| % | **Water** | 23 |
| | **Calories** | 55 |
| grams { | **Protein** | N.A. |
| | **Fat** | N.A. |
| | **Carbohydrate** | 14 |
| | **Calcium** | 35 |
| milligrams { | **Phosphorus** | 5 |
| | **Iron** | 2.6 |
| | **Sodium** | 19 |
| | **Potassium** | N.A. |
| | **Thiamin** | N.A. |
| | **Riboflavin** | .02 |
| | **Niacin** | Trace |
| | **Ascorbic Acid** | N.A. |
| unit | **Vitamin A** | N.A. |

# TABLE SYRUP BLENDS
## (CHIEFLY CORN SYRUP)

grams  **Weight:**  21

light & dark

| | | |
|---|---|---|
| % | Water | 24 |
| grams | Calories | 60 |
| | Protein | 0 |
| | Fat | 0 |
| | Carbohydrate | 15 |
| | Calcium | 9 |
| milligrams | Phosphorus | 3 |
| | Iron | .8 |
| | Sodium | *14 |
| | Potassium | 1 |
| | Thiamin | 0 |
| | Riboflavin | 0 |
| | Niacin | 0 |
| | Ascorbic Acid | 0 |
| unit | Vitamin A | 0 |

*Applies to product with added salt.

# WHITE SUGAR
## GRANULATED

| | grams | **Weight:** | 200 | 12 | 6 |
|---|---|---|---|---|---|

| | | white: granulated | white: granulated | white: granulated |
|---|---|---|---|---|
| % | **Water** | 1 | 1 | 1 |
| | **Calories** | 770 | 45 | 23 |
| grams | **Protein** | 0 | 0 | 0 |
| | **Fat** | 0 | 0 | 0 |
| | **Carbohydrate** | 199 | 12 | 7 |
| | **Calcium** | 0 | 0 | 0 |
| milligrams | **Phosphorus** | 0 | 0 | 0 |
| | **Iron** | .2 | Trace | Trace |
| | **Sodium** | 2 | Trace | Trace |
| | **Potassium** | 6 | Trace | Trace |
| | **Thiamin** | 0 | 0 | 0 |
| | **Riboflavin** | 0 | 0 | 0 |
| | **Niacin** | 0 | 0 | 0 |
| | **Ascorbic Acid** | 0 | 0 | 0 |
| unit | **Vitamin A** | 0 | 0 | 0 |

# WHITE SUGAR
## POWDERED

grams    **Weight:**    100

sifted,
spooned into
cup

| | | |
|---|---|---|
| % | **Water** | 1 |
| | **Calories** | 385 |
| grams { | **Protein** | 0 |
| | **Fat** | 0 |
| | **Carbohydrate** | 100 |
| | **Calcium** | 0 |
| milligrams { | **Phosphorus** | 0 |
| | **Iron** | .1 |
| | **Sodium** | 1 |
| | **Potassium** | 3 |
| | **Thiamin** | 0 |
| | **Riboflavin** | 0 |
| | **Niacin** | 0 |
| | **Ascorbic Acid** | 0 |
| unit | **Vitamin A** | 0 |

# JAMS, JELLIES AND GELATINS

JAMS AND PRESERVES
JELLIES
GELATIN, UNFLAVORED
GELATIN DESSERTS

# JAMS & PRESERVES

| | grams | Weight: | 20 | 14 |
|---|---|---|---|---|

| % | Water | 29 | 29 |
|---|---|---|---|
| | **Calories** | 55 | 40 |
| grams | Protein | Trace | Trace |
| | Fat | Trace | Trace |
| | Carbohydrate | 14 | 10 |
| | Calcium | 4 | 3 |
| milligrams | Phosphorus | 2 | 1 |
| | Iron | .2 | .1 |
| | Sodium | 2 | 2 |
| | Potassium | 15 | 12 |
| | Thiamin | Trace | Trace |
| | Riboflavin | .01 | Trace |
| | Niacin | Trace | Trace |
| | Ascorbic Acid | Trace | Trace |
| unit | Vitamin A | Trace | Trace |

# JELLIES

grams    **Weight:**        18              14

| | | | |
|---|---|---:|---:|
| % | **Water** | 29 | 29 |
| | **Calories** | 50 | 40 |
| grams | **Protein** | Trace | Trace |
| | **Fat** | Trace | Trace |
| | **Carbohydrate** | 13 | 10 |
| | **Calcium** | 4 | 3 |
| milligrams | **Phosphorus** | 1 | 1 |
| | **Iron** | .3 | .2 |
| | **Sodium** | 3 | 2 |
| | **Potassium** | 14 | 11 |
| | **Thiamin** | Trace | Trace |
| | **Riboflavin** | .01 | Trace |
| | **Niacin** | Trace | Trace |
| | **Ascorbic Acid** | 1 | 1 |
| unit | **Vitamin A** | Trace | Trace |

# GELATIN
## UNFLAVORED

grams | **Weight:** | 7

dry

| | | |
|---|---|---|
| % | **Water** | 13 |
| | **Calories** | 25 |
| grams | **Protein** | 6 |
| | **Fat** | Trace |
| | **Carbohydrate** | 0 |
| | **Calcium** | N.A. |
| | **Phosphorus** | N.A. |
| | **Iron** | N.A. |
| | **Sodium** | N.A. |
| milligrams | **Potassium** | N.A. |
| | **Thiamin** | N.A. |
| | **Riboflavin** | N.A. |
| | **Niacin** | N.A. |
| | **Ascorbic Acid** | N.A. |
| unit | **Vitamin A** | N.A. |

# GELATIN DESSERTS*

grams **Weight:** 240

| | | |
|---|---|---|
| % | **Water** | 84 |
| grams { | **Calories** | 140 |
| | **Protein** | 4 |
| | **Fat** | 0 |
| | **Carbohydrate** | 34 |
| | **Calcium** | N.A. |
| milligrams { | **Phosphorus** | N.A. |
| | **Iron** | N.A. |
| | **Sodium** | N.A. |
| | **Potassium** | N.A. |
| | **Thiamin** | N.A. |
| | **Riboflavin** | N.A. |
| | **Niacin** | N.A. |
| | **Ascorbic Acid** | N.A. |
| unit | **Vitamin A** | N.A. |

*Prepared with gelatin dessert powder and water.

# BEVERAGES

### ALCOHOLIC
BEER AND ALE
GIN, VODKA, RUM, WHISKEY
WINES, DESSERT AND TABLE

### CARBONATED, NON-ALCOHOLIC
CARBONATED WATER
COLA TYPE
CREAM SODA
FRUIT-FLAVORED SODAS
GINGER ALE
ROOT BEER

### MISCELLANEOUS
CHOCOLATE-FLAVORED BEVERAGE

SEE ALSO FRUIT JUICES

# BEER AND ALE

grams    **Weight:**              360

| | | |
|---|---|---|
| % | Water | 92 |
| | Calories | 150 |
| grams | Protein | 1 |
| | Fat | 0 |
| | Carbohydrate | 14 |
| | Calcium | 18 |
| | Phosphorus | 108 |
| | Iron | Trace |
| | Sodium | 25 |
| milligrams | Potassium | 90 |
| | Thiamin | .01 |
| | Riboflavin | .11 |
| | Niacin | 2.2 |
| | Ascorbic Acid | N.A. |
| unit | Vitamin A | N.A. |

# GIN, RUM,
## VODKA AND WHISKEY

| | | 80 proof | 86 proof | 90 proof |
|---|---|---|---|---|
| grams | Weight: | 42 | 42 | 42 |
| % | Water | 67 | 64 | 62 |
| grams | Calories | 95 | 105 | 110 |
| | Protein | N.A. | N.A. | N.A. |
| | Fat | N.A. | N.A. | N.A. |
| | Carbohydrate | Trace | Trace | Trace |
| | Calcium | N.A. | N.A. | N.A. |
| milligrams | Phosphorus | N.A. | N.A. | N.A. |
| | Iron | N.A. | N.A. | N.A. |
| | Sodium | Trace | Trace | Trace |
| | Potassium | 1 | 1 | 1 |
| | Thiamin | N.A. | N.A. | N.A. |
| | Riboflavin | N.A. | N.A. | N.A. |
| | Niacin | N.A. | N.A. | N.A. |
| | Ascorbic Acid | N.A. | N.A. | N.A. |
| unit | Vitamin A | N.A. | N.A. | N.A. |

# WINES
## DESSERT AND TABLE

| | | | |
|---|---|---|---|
| grams | **Weight:** | 103 | 102 |

| | | dessert | table |
|---|---|---|---|

| | | dessert | table |
|---|---|---|---|
| % | **Water** | 77 | 86 |
| | **Calories** | 140 | 85 |
| grams { | **Protein** | Trace | Trace |
| | **Fat** | 0 | 0 |
| | **Carbohydrate** | 8 | 4 |
| | **Calcium** | 8 | 9 |
| milligrams { | **Phosphorus** | N.A. | 10 |
| | **Iron** | N.A. | .4 |
| | **Sodium** | 4 | 5 |
| | **Potassium** | 77 | 94 |
| | **Thiamin** | .01 | Trace |
| | **Riboflavin** | .02 | .01 |
| | **Niacin** | .2 | .1 |
| | **Ascorbic Acid** | N.A. | N.A. |
| unit | **Vitamin A** | N.A. | N.A. |

# CARBONATED WATER

grams **Weight:** 366

| | | |
|---|---|---|
| % | **Water** | 92 |
| grams { | **Calories** | 115 |
| | **Protein** | 0 |
| | **Fat** | 0 |
| | **Carbohydrate** | 29 |
| | **Calcium** | N.A. |
| milligrams { | **Phosphorus** | N.A. |
| | **Iron** | N.A. |
| | **Sodium** | N.A. |
| | **Potassium** | N.A. |
| | **Thiamin** | 0 |
| | **Riboflavin** | 0 |
| | **Niacin** | 0 |
| | **Ascorbic Acid** | 0 |
| unit | **Vitamin A** | 0 |

# COLA
## (TYPE)

grams    **Weight:**          369

| % | Water | 90 | |
|---|---|---|---|
| | **Calories** | 145 | |
| | **Protein** | 0 | |
| | **Fat** | 0 | |
| | **Carbohydrate** | 37 | |
| | **Calcium** | N.A. | |
| | **Phosphorus** | N.A. | |
| | **Iron** | N.A. | |
| | **Sodium** | N.A. | |
| | **Potassium** | N.A. | |
| | **Thiamin** | 0 | |
| | **Riboflavin** | 0 | |
| | **Niacin** | 0 | |
| | **Ascorbic Acid** | 0 | |
| | **Vitamin A** | 0 | |

grams (Protein through Calcium), milligrams (Phosphorus through Vitamin A), unit

# CREAM SODA

grams **Weight:** 371

| | | |
|---|---|---|
| % | Water | 89 |
| grams | Calories | 160 |
| | Protein | 0 |
| | Fat | 0 |
| | Carbohydrate | 40.8 |
| | Calcium | N.A. |
| milligrams | Phosphorus | N.A. |
| | Iron | N.A. |
| | Sodium | N.A. |
| | Potassium | 0 |
| | Thiamin | 0 |
| | Riboflavin | 0 |
| | Niacin | 0 |
| | Ascorbic Acid | 0 |
| unit | Vitamin A | 0 |

# FRUIT FLAVORED SODAS
## (AND TOM COLLINS MIXER)

grams  **Weight:**  372

| | | |
|---|---|---|
| % | **Water** | 88 |
| | **Calories** | 170 |
| | **Protein** | 0 |
| grams | **Fat** | 0 |
| | **Carbohydrate** | 45 |
| | **Calcium** | N.A. |
| | **Phosphorus** | N.A. |
| | **Iron** | N.A. |
| milligrams | **Sodium** | N.A. |
| | **Potassium** | N.A. |
| | **Thiamin** | 0 |
| | **Riboflavin** | 0 |
| | **Niacin** | 0 |
| | **Ascorbic Acid** | 0 |
| unit | **Vitamin A** | 0 |

# GINGER ALE

grams **Weight:** 366

| | | |
|---|---|---|
| % | **Water** | 92 |
| | **Calories** | 115 |
| grams | **Protein** | 0 |
| | **Fat** | 0 |
| | **Carbohydrate** | 29 |
| | **Calcium** | N.A. |
| | **Phosphorus** | N.A. |
| | **Iron** | N.A. |
| milligrams | **Sodium** | N.A. |
| | **Potassium** | N.A. |
| | **Thiamin** | 0 |
| | **Riboflavin** | 0 |
| | **Niacin** | 0 |
| | **Ascorbic Acid** | 0 |
| unit | **Vitamin A** | 0, |

# ROOT BEER

| grams | **Weight:** | 370 |

| | | |
|---|---|---|
| % | **Water** | 90 |
| | **Calories** | 150 |
| grams | **Protein** | 0 |
| | **Fat** | 0 |
| | **Carbohydrate** | 39 |
| | **Calcium** | N.A. |
| milligrams | **Phosphorus** | N.A. |
| | **Iron** | N.A. |
| | **Sodium** | N.A. |
| | **Potassium** | N.A. |
| | **Thiamin** | 0 |
| | **Riboflavin** | 0 |
| | **Niacin** | 0 |
| | **Ascorbic Acid** | 0 |
| unit | **Vitamin A** | 0 |

# CHOCOLATE-FLAVORED BEVERAGE

| | Weight: | 28 | 28 |
|---|---|---|---|
| grams | | | |

| | | with nonfat dry milk | without milk |
|---|---|---|---|
| % | **Water** | 2 | 1 |
| | **Calories** | 100 | 100 |
| grams | **Protein** | 5 | 1 |
| | **Fat** | 1 | 1 |
| | **Carbohydrate** | 20 | 25 |
| | **Calcium** | 167 | 9 |
| milligrams | **Phosphorus** | 155 | 48 |
| | **Iron** | .5 | .6 |
| | **Sodium** | N.A. | N.A. |
| | **Potassium** | 227 | 142 |
| | **Thiamin** | .04 | .01 |
| | **Riboflavin** | .21 | .03 |
| | **Niacin** | .2 | .1 |
| | **Ascorbic Acid** | 1 | 0 |
| unit | **Vitamin A** | 10 | N.A. |

# MISCELLANEOUS

### BAKING POWDERS
### CIDER VINEGAR
### YEAST

# BAKING POWDERS
## FOR HOME USE

| grams | Weight: | 3.0 | 2.9 |
|---|---|---|---|
| | | *with mono-calcium phosphate monohydrate | *with mono-calcium phosphate monohydrate, calcium sulfate |
| % | **Water** | 2 | 1 |
| grams { | **Calories** | 5 | 5 |
| | **Protein** | Trace | Trace |
| | **Fat** | Trace | Trace |
| | **Carbohydrate** | 1 | 1 |
| milligrams { | **Calcium** | 58 | 183 |
| | **Phosphorus** | 87 | 45 |
| | **Iron** | N.A. | N.A. |
| | **Sodium** | 329 | 290 |
| | **Potassium** | 5 | N.A. |
| | **Thiamin** | 0 | 0 |
| | **Riboflavin** | 0 | 0 |
| | **Niacin** | 0 | 0 |
| | **Ascorbic Acid** | 0 | 0 |
| unit | **Vitamin A** | 0 | 0 |

*Sodium aluminum sulfate.

# BAKING POWDERS
## FOR HOME USE

| | grams | | |
|---|---|---|---|
| **Weight:** | | 3.0 | 2.9 |

| | | straight phosphate | low sodium |
|---|---|---|---|
| % | **Water** | 2 | 2 |
| | **Calories** | 5 | 5 |
| grams | **Protein** | Trace | Trace |
| | **Fat** | Trace | Trace |
| | **Carbohydrate** | 1 | 2 |
| | **Calcium** | 239 | 207 |
| milligrams | **Phosphorus** | 359 | 314 |
| | **Iron** | N.A. | N.A. |
| | **Sodium** | 312 | Trace |
| | **Potassium** | 6 | 471 |
| | **Thiamin** | 0 | 0 |
| | **Riboflavin** | 0 | 0 |
| | **Niacin** | 0 | 0 |
| | **Ascorbic Acid** | 0 | 0 |
| unit | **Vitamin A** | 0 | 0 |

503

# CIDER VINEGAR

grams **Weight:** 15

| % | Water | 94 |
|---|---|---|
| grams | Calories | Trace |
| | Protein | Trace |
| | Fat | 0 |
| | Carbohydrate | 1 |
| | Calcium | 1 |
| milligrams | Phosphorus | 1 |
| | Iron | .1 |
| | Sodium | Trace |
| | Potassium | 15 |
| | Thiamin | N.A. |
| | Riboflavin | N.A. |
| | Niacin | N.A. |
| | Ascorbic Acid | N.A. |
| unit | Vitamin A | N.A. |

# YEAST

| grams | Weight: | 7 | 8 |
|---|---|---|---|

|  | | Baker's dry, active | Brewer's dry |
|---|---|---|---|

| % | **Water** | 5 | 5 |
|---|---|---|---|
| | **Calories** | 20 | 25 |
| grams | **Protein** | 3 | 3 |
| | **Fat** | Trace | Trace |
| | **Carbohydrate** | 3 | 3 |
| | **Calcium** | 3 | *17 |
| | **Phosphorus** | 90 | 140 |
| | **Iron** | 1.1 | 1.4 |
| milligrams | **Sodium** | 4 | 10 |
| | **Potassium** | 140 | 152 |
| | **Thiamin** | .16 | 1.25 |
| | **Riboflavin** | .38 | .34 |
| | **Niacin** | 2.6 | 3.0 |
| | **Ascorbic Acid** | Trace | Trace |
| unit | **Vitamin A** | Trace | Trace |

*Value may vary from 6 to 60 mg.

# RECOMMENDED DAILY
# DIETARY ALLOWANCES
## (RDA'S)[1]

The RDA's are amounts of nutrients recommended by the Food and Nutrition Board of the National Research Council and are considered adequate for maintenance of good nutrition in healthy persons in the United States. The allowances are revised from time to time in accordance with newer knowledge of nutritional needs.

NOTE — They should not be confused with the U.S. Recommended Daily Allowances (U.S. RDA's). The U.S. RDA's are the amounts of protein, vitamins, and minerals established by the Food and Drug Administration as standards for nutrition labeling.

---

[1]Source: Adapted from Recommended Dietary Allowances, 8th ed., 1974, National Academy of Sciences.

# INFANTS

| Age: | 0-0.5 | 0.5-1 |
|---|---|---|
| **Weight** | | |
| Kilograms: | 6 | 9 |
| Pounds: | 14 | 20 |
| **Height:** | | |
| Centimeters: | 60 | 71 |
| Inches: | 24 | 28 |
| **Food Energy** | | |
| (Calories) | kg x 117 | kg x 108 |
| | lb x 53.2 | lb x 49.1 |
| **Protein** | | |
| (Grams) | kg x 2.2 | kg x 2.0 |
| | lb x 1.0 | lb x 0.9 |
| **Calcium** | | |
| (Milligrams) | 360 | 540 |
| **Phosphorus** | | |
| (Milligrams) | 240 | 400 |
| **Iron** | | |
| (Milligrams) | 10 | 15 |
| **Vitamin A** | | |
| (International Units) | 1,400 | 2,000 |
| **Thiamin** | | |
| (Milligrams) | 0.3 | 0.5 |
| **Riboflavin** | | |
| (Milligrams) | 0.4 | .6 |
| **Niacin** | | |
| (Milligrams) | 5 | 8 |
| **Ascorbic Acid** | | |
| (Milligrams) | 35 | 35 |

# CHILDREN

| Age: | 1-3 | 4-6 | 7-10 |
|---|---|---|---|
| **Weight** | | | |
| Kilograms: | 13 | 20 | 30 |
| Pounds: | 28 | 44 | 66 |
| **Height:** | | | |
| Centimeters: | 86 | 110 | 135 |
| Inches: | 34 | 44 | 54 |
| **Food Energy** | | | |
| (Calories) | 1,300 | 1,800 | 2,400 |
| **Protein** | | | |
| (Grams) | 23 | 30 | 36 |
| **Calcium** | | | |
| (Milligrams) | 800 | 800 | 800 |
| **Phosphorus** | | | |
| (Milligrams) | 240 | 400 | |
| **Iron** | | | |
| (Milligrams) | 10 | 15 | |
| **Vitamin A** | | | |
| (International Units) | 1,400 | 2,000 | |
| **Thiamin** | | | |
| (Milligrams) | 0.3 | 0.5 | |
| **Riboflavin** | | | |
| (Milligrams) | 0.4 | .6 | |
| **Niacin** | | | |
| (Milligrams) | 5 | 8 | |
| **Ascorbic Acid** | | | |
| (Milligrams) | 35 | 35 | |

# MALES

| Age: | 11-14 | 15-18 | 19-22 |
|---|---|---|---|
| **Weight** | | | |
| Kilograms: | 44 | 61 | 67 |
| Pounds: | 97 | 134 | 147 |
| **Height:** | | | |
| Centimeters: | 158 | 172 | 172 |
| Inches: | 63 | 69 | 69 |
| **Food Energy** | | | |
| (Calories) | 2,800 | 3,000 | 3,000 |
| **Protein** | | | |
| (Grams) | 44 | 54 | 54 |
| **Calcium** | | | |
| (Milligrams) | 1,200 | 1,200 | 800 |
| **Phosphorus** | | | |
| (Milligrams) | 1,200 | 1,200 | 800 |
| **Iron** | | | |
| (Milligrams) | 18 | 18 | 10 |
| **Vitamin A** | | | |
| (International Units) | 5,000 | 5,000 | 5,000 |
| **Thiamin** | | | |
| (Milligrams) | 1.4 | 1.5 | 1.5 |
| **Riboflavin** | | | |
| (Milligrams) | 1.5 | 1.8 | 1.8 |
| **Niacin** | | | |
| (Milligrams) | 18 | 20 | 20 |
| **Ascorbic Acid** | | | |
| (Milligrams) | 45 | 45 | 45 |

# MALES

| Age: | 23-50 | 50$ |
|---|---|---|
| **Weight** | | |
| Kilograms: | 70 | 70 |
| Pounds: | 154 | 154 |
| **Height:** | | |
| **Centimeters:** | 172 | 172 |
| Inches: | 69 | 69 |
| **Food Energy** | | |
| (Calories) | 2,700 | 2,400 |
| **Protein** | | |
| (Grams) | 56 | 56 |
| **Calcium** | | |
| (Milligrams) | 800 | 800 |
| **Phosphorus** | | |
| (Milligrams) | 800 | 800 |
| **Iron** | | |
| (Milligrams) | 10 | 10 |
| **Vitamin A** | | |
| (International Units) | 5,000 | 5,000 |
| **Thiamin** | | |
| (Milligrams) | 1.4 | 1.2 |
| **Riboflavin** | | |
| (Milligrams) | 1.6 | 1.5 |
| **Niacin** | | |
| (Milligrams) | 18 | 16 |
| **Ascorbic Acid** | | |
| (Milligrams) | 45 | 45 |

# FEMALES

| Age: | 11-14 | 15-18 | 19-22 |
|---|---|---|---|
| **Weight** | | | |
| Kilograms: | 44 | 54 | 58 |
| Pounds: | 97 | 119 | 128 |
| **Height:** | | | |
| Centimeters: | 155 | 162 | 162 |
| Inches: | 62 | 65 | 65 |
| **Food Energy** | | | |
| (Calories) | 2,400 | 2,100 | 2,100 |
| **Protein** | | | |
| (Grams) | 44 | 48 | 46 |
| **Calcium** | | | |
| (Milligrams) | 1,200 | 1,200 | 800 |
| **Phosphorus** | | | |
| (Milligrams) | 1,200 | 1,200 | 800 |
| **Iron** | | | |
| (Milligrams) | 18 | 18 | 18 |
| **Vitamin A** | | | |
| (International Units) | 4,000 | 4,000 | 4,000 |
| **Thiamin** | | | |
| (Milligrams) | 1.2 | 1.1 | 1.1 |
| **Riboflavin** | | | |
| (Milligrams) | 1.3 | 1.4 | 1.4 |
| **Niacin** | | | |
| (Milligrams) | 16 | 14 | 14 |
| **Ascorbic Acid** | | | |
| (Milligrams) | 45 | 45 | 45 |

# FEMALES

| Age: | 23-50 | 50$ |
|---|---|---|
| **Weight** | | |
| Kilograms: | 58 | 58 |
| Pounds: | 128 | 128 |
| **Height:** | | |
| Centimeters: | 162 | 162 |
| Inches: | 65 | 65 |
| **Food Energy** | | |
| (Calories) | 2,000 | 1,800 |
| **Protein** | | |
| (Grams) | 46 | 46 |
| **Calcium** | | |
| (Milligrams) | 800 | 800 |
| **Phosphorus** | | |
| (Milligrams) | 800 | 800 |
| **Iron** | | |
| (Milligrams) | 18 | 10 |
| **Vitamin A** | | |
| (International Units) | 4,000 | 4,000 |
| **Thiamin** | | |
| (Milligrams) | 1.0 | 1.0 |
| **Riboflavin** | | |
| (Milligrams) | 1.2 | 1.2 |
| **Niacin** | | |
| (Milligrams) | 13 | 12 |
| **Ascorbic Acid** | | |
| (Milligrams) | 45 | 45 |

# FEMALES

|  | Pregnant | Lactating |
|---|---|---|
| **Age:** | | |
| **Weight** | | |
| Kilograms: | | |
| Pounds: | | |
| **Height:** | | |
| Centimeters: | | |
| Inches: | | |
| **Food Energy** | | |
| (Calories) | +300 | +500 |
| **Protein** | | |
| (Grams) | **+30** | +20 |
| **Calcium** | | |
| (Milligrams) | 1,200 | 1,200 |
| **Phosphorus** | | |
| (Milligrams) | 1,200 | 1,200 |
| **Iron** | | |
| (Milligrams) | +18 | 18 |
| **Vitamin A** | | |
| (International Units) | 5,000 | 6,000 |
| **Thiamin** | | |
| (Milligrams) | +.3 | +.3 |
| **Riboflavin** | | |
| (Milligrams) | +.3 | +.5 |
| **Niacin** | | |
| (Milligrams) | +2 | +4 |
| **Ascorbic Acid** | | |
| (Milligrams) | 60 | 80 |

# ADDITIONAL
# NUTRITION SOURCES
## IN FOODS

### Vitamin B$_6$

Bananas

Whole-grain cereals

Chicken

Dry legumes

Egg yolk

Most dark-green
leafy vegetables

Most fish and shellfish

Muscle meats, liver and
kidney

Peanuts, peanut butter,
walnuts, and filberts

Potatoes and
sweet potatoes

Prunes and raisins

Yeast

### Vitamin B$_{12}$
(present in foods of animal origin only)

Kidney
Liver
Meat
Milk

Most cheese
Most fish
Shellfish
Whole egg and egg yolk

### Vitamin D

Vitamin D milks
Egg yolk

Saltwater fish
Liver

## VITAMIN E

Vegetable oils

Margarine

Whole-grain cereals

Peanuts

## FOLACIN

Liver
Dark-green vegetables
Dry beans

Peanuts
Wheat germ

## IODINE

Iodized salt

Seafood

## MAGNESIUM

Bananas
Whole-grain cereals
Dry beans
Milk

Most dark-green vegetables
Nuts
Peanuts and peanut butter

## ZINC

Shellfish
Meat
Cheese
Whole-grain cereals

Dry beans
Cocoa
Nuts

# How's Your Health?

Bantam publishes a line of informative books, written by top experts to help you toward a healthier and happier life.

| | | |
|---|---|---|
| ☐ 22660 | **VITAMIN BIBLE FOR YOUR KIDS** E. Mindell | $3.95 |
| ☐ 01376 | **THE FOOD ADDITIVES BOOK**<br>Nicholas Freydberg & Willis A. Gortnor<br>**A Large Format Book** | $9.95 |
| ☐ 20830 | **ALTERNATIVE APPROACH TO ALLERGIES**<br>T. Randolph & R. Moss, Ph.D. | $3.95 |
| ☐ 20562 | **SECOND OPINION** Isadore Rosenfeld | $3.95 |
| ☐ 20669 | **HEART RISK BOOK** Aram V. Chobanian, M.D. &<br>Lorraine Loviglio | $2.50 |
| ☐ 20494 | **THE PRITIKIN PERMANENT WEIGHT-LOSS<br>MANUAL** Nathan Pritikin | $3.95 |
| ☐ 20322 | **HEALTH FOR THE WHOLE PERSON:**<br>**The Complete Guide to Holistic Health**<br>Hastings, et. al. | $3.95 |
| ☐ 20163 | **PREVENTION OF ALCOHOLISM THROUGH<br>NUTRITION** Roger Williams | $2.95 |
| ☐ 23519 | **THE BRAND NAME NUTRITION COUNTER**<br>Jean Carper | $2.95 |
| ☐ 23066 | **NUTRITION AGAINST DISEASE** Roger T. Williams | $3.95 |
| ☐ 22801 | **HOW TO LIVE 365 DAYS A YEAR THE<br>SALT FREE WAY** Brunswick, Love & Weinberg | $3.50 |
| ☐ 20621 | **THE PILL BOOK** Simon & Silverman | $3.95 |
| ☐ 20134 | **PRITIKIN PROGRAM FOR DIET AND EXERCISE**<br>Pritikin & McGrady, Jr. | $3.95 |
| ☐ 20052 | **THE GIFT OF HEALTH: A New Approach to<br>Higher Quality, Lower Cost Health Care**<br>Shames & Shames | $2.95 |
| ☐ 23147 | **WHICH VITAMINS DO YOU NEED?** Martin Ebon | $3.50 |
| ☐ 23251 | **PSYCHODIETETICS** Cheraskin, et. al. | $2.95 |

**Buy them at your local bookstore or use this handy coupon for ordering:**

---

Bantam Books, Inc., Dept. HN, 414 East Golf Road, Des Plaines, Ill. 60016

Please send me the books I have checked above. I am enclosing $_____
(please add $1.25 to cover postage and handling). Send check or money order
—no cash or C.O.D.'s please.

Mr/Mrs/Miss_____

Address_____

City_____State/Zip_____

HN—3/83

Please allow four to six weeks for delivery. This offer expires 9/83.

# NEED MORE INFORMATION ON YOUR HEALTH AND NUTRITION?

Read the books that will lead you to
a happier and healthier life.

| | | |
|---|---|---|
| ☐ 20925 | WHAT'S IN WHAT YOU EAT<br>Will Eisner | $3.95 |
| ☐ 20175 | A DICTIONARY OF SYMPTOMS<br>Dr. J. Gomez | $3.95 |
| ☐ 01464 | MY BODY, MY HEALTH<br>Stewart & Hatcher<br>(A large Format Book) | $10.95 |
| ☐ 22872 | WOMEN AND THE CRISES<br>IN SEX HORMONES<br>G. Seamans | $4.50 |
| ☐ 20939 | PRE-MENSTRUAL TENSION<br>J. Lever with Dr. M. Brush | $2.95 |
| ☐ 22537 | HEALING FROM WITHIN<br>D. Jaffe | $3.50 |
| ☐ 22673 | CONCISE MEDICAL DICTIONARY<br>Laurance Urdange Associates | $4.95 |
| ☐ 23335 | GH-3-WILL IT KEEP YOU YOUNG<br>LONGER<br>H. Bailey | $3.95 |
| ☐ 22976 | THE HERB BOOK<br>J. Lust | $4.50 |
| ☐ 22571 | HIGH LEVEL WELLNESS<br>D. Ardell | $3.25 |
| ☐ 23031 | HOPE AND HELP FOR YOUR NERVES<br>C. Weekes | $3.50 |
| ☐ 22503 | PEACE FROM NERVOUS SUFFERING<br>C. Weekes | $3.50 |
| ☐ 20975 | SIMPLE, EFFECTIVE TREATMENT OF<br>AGORAPHOBIA<br>C. Weekes | $3.50 |
| ☐ 20806 | SPIRULINA: WHOLE FOOD REVOLUTION<br>L. Switzer | $2.95 |

Bantam Books, Inc., Dept. HN2, 414 East Golf Road, Des Plaines, Ill. 60016

Please send me the books I have checked above. I am enclosing $_____
(please add $1.25 to cover postage and handling). Send check or money
order—no cash or C.O.D.'s please.

Mr/Mrs/Miss _____

Address _____

City _____ State/Zip _____

HN2—3/83
Please allow four to six weeks for delivery. This offer expires 9/83.

SAVE $2.00 ON YOUR NEXT BOOK ORDER!

# BANTAM BOOKS 🐓

*Shop-at-Home* ——
*Catalog*

Now you can have a complete, up-to-date catalog of Bantam's inventory of over 1,600 titles—including hard-to-find books. And, you can save $2.00 on your next order by taking advantage of the money-saving coupon you'll find in this illustrated catalog. Choose from fiction and non-fiction titles, including mysteries, historical novels, westerns, cookbooks, romances, biographies, family living, health, and more. You'll find a description of most titles. Arranged by categoreis, the catalog makes it easy to find your favorite books and authors and to discover new ones.

So don't delay—send for this shop-at-home catalog and save money on your next book order.

Just send us your name and address and 50¢ to defray postage and handling costs.

---

**BANTAM BOOKS, INC.**
**Dept. FC, 414 East Golf Road, Des Plaines, Ill. 60016**

Mr./Mrs./Miss _____
(please print)

Address _____

City _____ State _____ Zip _____

Do you know someone who enjoys books? Just give us their names and addresses and we'll send them a catalog too at no extra cost!

Mr./Mrs./Miss _____

Address _____

City _____ State _____ Zip _____

Mr./Mrs./Miss _____

Address _____

City _____ State _____ Zip _____

FC—2/83